Country Capitalism

Country Capitalism

HOW CORPORATIONS FROM THE AMERICAN SOUTH REMADE OUR ECONOMY AND THE PLANET

Bart Elmore

FF
FF

A Ferris and Ferris Book

THE UNIVERSITY OF NORTH CAROLINA PRESS

Chapel Hill

This book was published under the Marcie Cohen Ferris and William R. Ferris Imprint of the University of North Carolina Press.

Manufactured in the United States of America

LIBRARY OF CONGRESS CATALOGING-IN-PUBLICATION DATA
Names: Elmore, Bartow J., author.
Title: Country capitalism : how corporations from the American South remade our economy and the planet / Bart Elmore.
Description: Chapel Hill : University of North Carolina Press, [2023] | "A Ferris and Ferris book." | Includes bibliographical references and index.
Identifiers: LCCN 2022053867 | ISBN 9781469673332 (cloth ; alk. paper) | ISBN 9781469673349 (ebook)
Subjects: LCSH: Coca-Cola Company. | Delta Air Lines. | Wal-Mart (Firm) | FedEx Corporation. | Bank of America. | International business enterprises—Southern States—History—20th century. | International business enterprises—Environmental aspects—Southern States. | Capitalism—Southern States—History—20th century. | Business logistics—Environmental aspects. | Globalization—Environmental aspects. | Global warming—Economic aspects. | Sustainable development. | BISAC: HISTORY / United States / State & Local / South (AL, AR, FL, GA, KY, LA, MS, NC, SC, TN, VA, WV) | SCIENCE / Environmental Science (see also Chemistry / Environmental)
Classification: LCC HD2798.S68 E566 2023 | DDC 338.70975—dc23/eng/20230117
LC record available at https://lccn.loc.gov/2022053867

For River and Blue

Contents

Preface

Richard Smith, the son of FedEx founder Fred Smith and the head of FedEx's Americas division, had just been through quite a year. It was March 2021, and twelve months prior, the world had spiraled into chaos as the COVID-19 pandemic swept the globe. While many Americans were quietly going about their usual routines, oblivious to the severity of what was about to befall them, Smith and others at FedEx were watching a rapidly replicating strand of RNA spread in vectors that aligned with the company's international network. Smith's company operated in over 220 countries and territories, including Wuhan, China, and the firm had helped get personal protective equipment to Asia in the first days of the virus outbreak. Months later, when the scale of the crisis had become unmistakable, federal officials called Smith and executives from FedEx's corporate rival UPS to Washington to figure out the logistics of getting newly developed vaccines into Americans' arms.[1]

Smith, an energetic, gray-haired man in his forties who once played football at the University of Virginia, was amped up when talking about FedEx's role in the coronavirus vaccine rollout. FedEx was a family business, a source of great pride. He had grown up in Memphis, Tennessee, and watched his father build an air express delivery firm from the ground up in the 1970s and 1980s. He remembered tagging along with his dad as a kid to FedEx's "Hub," the company headquarters located in the Memphis International Airport. Cargo planes came in from all over the country, arriving around 11 P.M. with all sorts of urgent packages that needed to be delivered to towns thousands of miles away. Overnight, FedEx employees—many of them college students who worked part-time—would spend the next few hours in a frenzied rush to move packages onto outbound flights. A massive digital clock loomed over workers in the Hub, and as seconds ticked down, suspense grew. All planes had to be ready to go by morning. That was the FedEx system, the hub-and-spoke delivery network that revolutionized air express delivery and, with it, our global economy.[2]

As Smith recounted the story of how FedEx got involved in COVID-19 vaccine delivery, he spoke with a kind of boyish sense of awe, one he probably experienced on those first visits to the Hub. He said the company had been preparing for a moment like the pandemic for fifty years. "This is what Federal Express, the original integrated air-ground system that connects all those origin and destination pairs in a highly efficient hub-and-spoke system, was designed to do," Smith said.[3]

In 2009, the H1N1 swine flu outbreak spurred FedEx to ramp up investments in cold-chain technology that would enable the firm to safely store and distribute vaccines and other biotechnology products at subzero temperatures. FedEx founder Fred Smith, Richard's dad, called the H1N1 scare a "wake-up call": he began sinking even more capital in broadening the company's cold-chain infrastructure and pharmaceutical logistics capacity. By the end of 2009, the firm announced the creation of SenseAware, a new, patented technology the size of an iPhone that could be added to a time-sensitive and temperature-controlled package. The tech would allow FedEx trackers to get real-time delivery information throughout its global network. It was an amazing device, allowing FedEx Priority Alert agents—dubbed "guardian angels" by Richard Smith—to track everything from the barometric pressure in a package to the number of times a parcel was accidentally dropped during its journey. For particularly sensitive deliveries, SenseAware could issue alerts if a package breached a geo-fence established by the shipper.[4]

It took roughly a decade for Richard Smith's team to develop a lightweight Bluetooth device the size of a key fob, known as SenseAware ID, that could be added to a wider array of packages. By the spring of 2020, Smith also announced a new collaboration with Microsoft known as FedEx Surround, a data system that provides "near-real-time analytics into shipment tracking." With Surround, a FedEx courier in Arkansas shipping a package destined for California could learn about a storm due to hit Denver in a few hours and consequently initiate a package reroute that would avoid the inclement weather. This was a remarkable step forward in logistics, one that attracted the attention of the federal government when the COVID crisis spread in 2020.[5]

FedEx joined UPS as one of the two core distributors of coronavirus vaccines in the United States. By May 2021, FedEx had delivered roughly 100 million vaccines throughout all fifty states and U.S. territories, including to Virginia, where my family and I were waiting for relief from the pandemic.[6]

For months we had been donning masks as our only defense against the coronavirus, masks that might have been shipped at one point by FedEx. After all, the company had played a major role in Project Airbridge, delivering fifty-five kilotons of personal protection equipment and roughly 2 billion face masks over the course of the pandemic.[7] In short, for the past year, the infrastructure Richard Smith had put in place had helped keep me and my loved ones safe from a deadly virus.

To be honest, if I had not been writing this book while the coronavirus chaos was unfolding, I probably would not have thought much about FedEx's role in ameliorating the worst problems of a global pandemic. The story of the race to manufacture vaccines—a story that centered on complicated genetic engineering techniques deployed in the laboratories of Pfizer, Moderna, and Johnson & Johnson—was in many ways more captivating.

But for the past several years, I had been thinking a lot about companies from the American South that created systems to make things like rapid vaccine rollouts possible. I came to see the critical role southern firms played in creating infrastructure that allowed goods to travel rapidly from one remote market of the world to another. Businesses like Memphis-born FedEx were commercial blood thinners, accelerating the flow of goods through our global economy.

That acceleration had the potential to save lives, as FedEx's COVID response revealed, but it also could do harm. While some businesses suffered losses during the pandemic, FedEx enjoyed considerable growth, in part because people needed more goods delivered directly to their door in a time of social distancing. But that growth brought with it a bigger environmental footprint. In sustainability reports, FedEx lauded the fact that it was seeing reductions in its "emissions intensity" throughout the pandemic, but that metric measured its greenhouse gas emissions in proportion to the revenues it was generating. Looking at its total carbon footprint, it is clear that the company was producing more greenhouse gases over time in its corporate system, not less. Between 2019 and the end of 2021, the firm's total greenhouse gas emissions, which exceeded 21 million tons of CO_2 equivalents in 2021, increased by nearly 10 percent. This was bad news at a time when the world faced a worsening climate change crisis.[8]

This book, then, is an attempt to explain how it is that I can push a button right now on my computer that will bring goods—some life-saving—from thousands of miles away to my doorstep in a matter of minutes. It is also a book that explores the ecological consequences of that button pushing. It is a history that tries to explain both the profits and perils of our express-delivery, fly-by-night, buy-on-credit economy, one in which I am admittedly very much embedded.

Columbus, Ohio
July 2022

Country Capitalism

Introduction

THE AMERICAN SOUTH AND THE RURAL ROADS TO OUR EARTH-CHANGING ECONOMY

As these words hit the page, I can see my eighteen-year-old self clearly: a scrawny, six-foot-something Dartmouth freshman from Atlanta, Georgia, sitting on a hillside, awaiting a week of orientation in the mountains of New Hampshire. Pictures from that week are revealing. The pent-up anxiety is nearly palpable, though smiles try to hide the fear. I was roughly 1,100 miles away from home and wondering, probably like many of my soon-to-be college friends, whether people would accept me in this new world I found myself in.

For the next four years, Hanover, New Hampshire, became a place where I questioned everything I knew about my past. I began to try to figure out who I was, and part of that journey included figuring out what it meant to be a "southerner" living in New England. I'm not obsessed with this question now, but back then it was something I thought about a lot. That's because I was able to see my home region for the first time through the eyes of people who had not grown up where I had.

In those early days in college, I was teased about being from a backward place and was even told the American South was the land of the unenlightened. These gibes were often annoying, but I understood where they came from. My home state had a lot of problems. Persistent racial segregation, extreme economic inequality, inadequate access to good public schools and health care facilities for poor and minority citizens—these were just some of the markers of injustice that defined the region I grew up in, markers I came to see more clearly from a distance.

In the many conversations I had in school, people always gave me an out: "Well, you're from Atlanta" was the familiar quip, often delivered as a way of saying, "Well, you're not really from *the* South." Even after college, this was the thing people said when they were trying to figure out why I didn't have a thick southern accent (though many people in my family did) or why I might promote environmentalism, vegetarian eating, or other customs that seemed at odds with the traditions of the American South. "Well, you're from Atlanta." I understood what they meant.

I even sometimes bought into the rhetoric. I remember trying to convince my friends that I was not from a backward place by ticking off a list of all the big businesses that hailed from my hometown. Think about Coca-Cola, I would boast: this company, one of the best-known global brands, started in a small pharmacy in downtown Atlanta. Or Delta, the world's largest airline, whose home hub is Hartsfield-Jackson Atlanta International Airport. Or CNN, a world leader in news, which for decades broadcast from headquarters located just a short drive from my childhood home.

To sell the South to my friends, in other words, I was treating the booming businesses in my hometown as if they were proxies for progress.

My logic was deeply flawed, as I came to understand during my graduate work in southern history at the University of Virginia, where mentors reintroduced me to the region I called home. The thing I began to recognize more clearly in those first few years of graduate school was that economic progress in the American South was often deeply entwined with some of the most retrograde social institutions the world has ever seen. It was not, in other words, the South's lack of economic development that explained the region's ills. Rather, the horror of Jim Crow segregation was overlaid and embedded in the most advanced technologies of the day. Railroads and streetcars—the best inventions of the nineteenth century—often were the spaces where white supremacists fretted most when thinking about how to codify the separation of Blacks and whites in the American South. Country stores, which offered many Gilded Age southerners their first taste of new brands such as Coca-Cola, also became the kind of public spaces that demanded Jim Crow segregationists' close attention, precisely because these commercial centers were so new.[1] Children in the late nineteenth-century South sometimes experienced the horrors of Black lynchings while using brand-new Edison listening machines that contained recordings of these spectacle killings.[2]

In graduate school, I read the work of historians who showed how southern politicians and business boosters in the latter half of the twentieth century used tax abatements, municipal subsidies, "right-to-work" laws, and lax environmental regulations to entice new businesses and industries to the American South while simultaneously ignoring the social and ecological welfare of poor southerners, many of whom were minorities. Many Black communities remained impoverished despite the arrival of new factories and defense plants to the region in the post–World War II era. During the Cold War, southern leaders continued to pursue policies that were about economic growth for the region's "haves" while spurning tax revenue that could be channeled toward support and opportunity for the "have-nots."[3]

This is not to say that there has been no real progress in the American South in recent decades. Federal laws spurred by the activism of the modern civil rights movement did bring about an improvement in economic well-being for many white and Black southerners. Personal income, retail sales, and other business investments rose sharply in the American South after civil rights activists won some of their most important victories in the mid-1960s. These changes did not occur because southern business leaders openly embraced desegregation. Rather, as one southern historian put it, "southern business was compelled or at least impelled by outside authority to take actions serving its own best economic interest" in the years after the 1960s.[4]

I grew up in those boom years, where southern cities like Atlanta and Charlotte exploded with new skyscrapers, fancy restaurants, and big sports franchises. On the surface, it seemed like so much was going right.

Nevertheless, racial and social inequalities born of the Jim Crow era persisted in the decades following the civil rights movement. The southern economy may have boomed after the 1960s, but because southern state governments remained more committed to investing in firms instead of families, the sins of the Jim Crow era were never fully absolved.

And those sins included crimes against nature. Over the past several decades, a talented cadre of southern environmental historians has taken us through the fields, forests, mountains, and streams of the American South's ecological history. Brilliant works have helped us see how plantation agriculture and sharecropping exhausted the southern land and contributed to the impoverishment of poor Black and white citizens. Some scholars have cataloged the ecological degradation wrought by timber companies and mining firms that descended on the pine forests of the Piedmont and the mountains of Appalachia; others have detailed the avarice of unscrupulous developers who built risky properties in wetland ecosystems. Histories also catalog the human and ecological costs of chicken farming and chemical manufacture in the American South's dirtiest factories and plants.[5]

These stories of the southern environment have focused mainly on ecological harm to southern ecosystems within southern borders, but as I looked more closely into the history of southern business and the environment in my studies, I came to see that scholars were only telling part of the story. My main intervention in this book is to turn our gaze outward, to see how firms and institutions born in the American South have reshaped global environments far beyond their home region. As we'll see, the Southland has been more than just the site of ecological degradation and natural resource extraction. In many ways, it has also been an exporter, rather than just an importer, of ecological problems.[6]

I no longer point to big business growth in my hometown as a marker of progress as I once did when I was a college freshman so many years ago, but I've never fully shaken my interest in these enterprises either, in part because I believe that understanding southern business is critical to understanding our global economy and the ecosystems on which we all depend.

In a way, then, this book represents a return to two seemingly simple questions that have captured my interest for many years: How did so many businesses from the American South, many founded in the Jim Crow era, become so ubiquitous in markets across the globe? And as they grew to become global brands, what were the ecological costs?

To answer these questions, I've examined five southern companies: Coca-Cola, Delta Air Lines, Walmart, FedEx, and Bank of America. With each firm, I wanted to do two things. First, I sought to understand how the unique ecosystems of the American South gave rise to these businesses that had such outsized influence on our planet. How did the particularities of the southern environment, in other words, shape the rise of these multinational enterprises? In the book, I often use the term "commercial ecology" to describe the habitat in which these firms operated, and I do so intentionally to emphasize the complex interplay between environmental, political, and socioeconomic factors that shaped corporate strategies.[7]

Second, I wanted to know how these firms' particular operations affected global ecosystems well beyond southern borders. In other words, what were the national and international implications of logistics systems and business strategies developed in the unique commercial ecology of the American South?

The answer to the first question about the attributes of the southern environment that shaped the South's most successful firms proved surprising. I found that each of the five companies grew big by engaging in what I call "country capitalism," developing systems of servicing the rural countryside or smaller communities removed from major metropolitan centers. None of these businesses were primarily engaged in the kinds of extractive industries (timber, textiles, tobacco, cotton, and the like) scholars typically discuss when focusing on the economy and environmental history of the American South. Rather, all of these firms were primarily interested in becoming conduits of capitalism, perfecting methods of channeling secret ingredients, high-flying entrepreneurs, chemicals, dry goods, and investment capital—often originating in countries well beyond the Southland—to consumers and markets in remote corners of the globe.

The American South, of course, was not the only place where firms grew big by finding ways to channel goods and people to and from remote places. Companies such as UPS founded in Seattle (though now based in Atlanta) and Sears in Chicago also grew large in the early twentieth century by figuring out how to service rural and countryside markets. Environmental historians have written a great deal about how the Windy City was a kind of "nature's metropolis" in part because of its deep market connections to countryside communities that helped make the city a financial and commercial center in the Gilded Age.[8] And as we'll see in the story of Bank of America, California shared similarities with North Carolina when it came to developing banking systems that served rural communities.

But when we examine these five southern firms together, it becomes clear that the American South was also a place where some of the basic logistics systems that drive our economy today were perfected, and I contend that this was in part because of southern businesses' interest in servicing a vast southern countryside. As late as 1939, roughly two-thirds of all southerners still worked farmland or lived in communities that were smaller than 2,500 people.[9] This was roughly two decades after the U.S. Census Bureau reported for the first time, in 1920, that more Americans lived in cities than in rural communities.[10] The American South, in other words, remained more rural than many other parts of the country well into the mid-twentieth century. After World War II, New Deal policies transformed southern labor markets, and the civil rights movement helped reverse migration patterns and spurred new investments in southern cities. Later chapters examine this urbanization and suburbanization of the southern landscape, acknowledging that the American South was never static, always evolving, always in motion.[11] Yet through all these changes, the southern countryside and its more remote markets remained of interest to firms that had matured at a time when the American South was more rural and less urbanized. As a result, some of the American South's most successful firms looked upon rurality as an asset rather than a liability. Their businesses were about channeling capital, goods, and people through the economy, as opposed to directly extracting natural resources from the Southland and shipping those resources around the world. As they perfected their logistics systems, flight patterns, and financial arrangements, these southern firms changed their home region, the nation, and ultimately the globe.

It is worth acknowledging that the words "rural" and "countryside" are contested terms, and in this book, I acknowledge that the boundaries between what we might call rural, suburban, exurban, and metropolitan are blurred at best.[12] For example, in the 2010s the U.S. Census Bureau defined "urbanized areas" as communities with more than 50,000 people, and it also had a category for "urban clusters," areas that had a "population of at least 2,500 and less than 50,000." But back in the late nineteenth and early twentieth centuries, definitions were different: the Census Bureau labeled communities "towns" or "urban cities" if they had just 4,000 or 8,000 residents.[13] I am less interested in settling the debate about what precisely defines an urban or rural area than I am in acknowledging that southern businesses in this book saw value in less urbanized, more rural parts of the American South, and as a result, they developed unique logistics systems and business operations to serve these places, systems that have become critical to the functioning of our modern economy.[14]

Which brings me to the second major discovery I made in this work: that southern firms' emphasis on making money by swiftly channeling goods, people, and capital over vast geographic distances, often to more rural and remote corners of the world, led these firms toward an outsized ecological footprint on the whole globe. The conduits of capitalism featured in this book may not have owned the widest array of manufacturing facilities or polluting plants, but they nevertheless were the architects of commercial arteries that have accelerated the flow of goods and people through the economy over the past several decades. By the twenty-first century, these businesses committed themselves to addressing their environmental impacts, but they focused mainly on doing their job more efficiently—cutting out waste, investing in energy-efficient equipment, and so on—which in many cases increased their capacity to ship more goods through their corporate systems, with the attendant result of enormous greenhouse gas emissions and other major pollution problems.

I hope those reading this book, as well as these firms and the government agencies that regulate them, will think about the past as they consider future sustainability objectives and will recognize that corporations' single-minded obsession of designing ever-quicker, lightning-fast delivery systems is leading us toward a hotter planet, one that can't possibly sustain life on Earth as we know it if we don't change course.

The rural roads that led to our have-it-now, fly-by-night, buy-on-credit economy ran through the American South, though this is not how most people see it. Today, Jeff Bezos's Seattle-based firm Amazon gets most of the credit for creating the rapid channels of commercial exchange that pulse goods at lightning speed across the globe. Yet the so-called Amazon economy ultimately was built on top of logistics technology and sales philosophies that were in part developed by southern firms featured in this book. It was Coca-Cola's Robert Woodruff, a former truck salesman, who in the early twentieth century articulated his hope that his beverages would be "within arm's reach of desire," a mantra that so accurately sums up consumer expectations in our "Amazon economy" today. In the 1960s, Walmart's Sam Walton pioneered new advancements in the trucking industry, developing transportation networks, distribution hubs, and retail satellite systems that served the rural countryside and became the envy of all in the retail industry. Though Amazon does not have the same brick-and-mortar stores Walmart does, and though Bezos's company has had to develop unique logistics technology to distribute goods directly to customers' doors, Amazon's fulfillment centers, which bring goods in from a host of manufacturers and then ship them out to consumers, draw on supplier collaboration strategies and data-sharing techniques that Walmart pioneered so many years ago.[15] And it was Delta Air Lines in the 1950s that first developed the "hub-and-spoke" feeder system that brought passengers from small towns to Atlanta and then onward to global destinations, a system that FedEx built on to revolutionize the air express delivery industry in the 1970s, forcing firms like UPS—and, yes, ultimately Amazon—to invest in an advanced fleet of cargo aircraft capable of delivering packages swiftly and often overnight from one market to another.[16]

And while Charlotte-based Bank of America may seem like an odd fit for this mix of businesses that influenced Amazon's success, it turns out that financial firms are indeed part of the story because they were so central to expanding consumer credit in the twentieth century, credit that ultimately enabled Amazon purchases of just about everything. Jeff Bezos acknowledged this fact directly in 2019, saying that the credit card, first popularized by Bank of America, was one of the key pieces of Amazon's infrastructure that was "already in place" when Amazon emerged in 1994.[17] And as this book went to press, Amazon partnered with Bank of America to begin offering small loans to small businesses that might become suppliers to the firm. In 2021, Amazon's interest in financing had *Forbes* asking the question, "Is Amazon building the next-generation bank?" Clearly, in other words, Amazon realized that Bank of America's success at expanding business and consumer credit, success that first began in the countryside in California and North Carolina, was a central ingredient in its business operations.[18]

Which is why if we want to understand the roots of the "Amazon economy" with an eye toward creating a more eco-conscious global network of exchange, we must study the conduits of capitalism that emerged in the American South long before Jeff Bezos conceived of his multibillion-dollar business.

So it is that we travel back in time, to a segregated southern city that had not yet heard of diesel-fueled eighteen-wheelers or jet-fuel-burning airliners. There, a pharmacist named Asa Candler was about to build one of the biggest brands the world has ever known.

PART ONE

I'd Like to Buy the World a Coke

Chapter One

I THOUGHT IT WOULD BE PROFITABLE IN THE COUNTRY IF I COULD ONLY GET IT THERE

A young Asa Candler had been wanting to leave the countryside in search of opportunity in the city. Born in 1851, Candler grew up on his family's farm in Villa Rica, Georgia, a small town approximately thirty-six miles west of Atlanta founded in the wake of the Georgia gold rush. His father, Samuel, was a rich man, having made enough money as a planter and general-store owner to purchase and enslave nearly twenty Black laborers by 1860. This made him one of Carroll County's largest slave owners. But Samuel's reliance on inhuman bondage to bring profits from the land did not prevent him from putting his own children to work. Samuel and his wife Martha preached Protestant discipline to their children, including the familiar idea that the devil made use of idle hands. Spring, summer, and fall, Asa toiled in the hot Georgia sun, cultivating corn and cotton, and weathering the bruises associated with field work. (Family lore has it that he even lost hearing in one of his ears after falling off a farm wagon.)[1] When Asa was ten, the Civil War forced his father to shut down his merchant operations. The plantation became the family's only means of income. Asa's son Howard recalled that during this time his grandfather Samuel "spent most of his time on the farm and exerted a much more direct influence on Father's developing character than before."[2]

But Candler was clearly interested more in commerce than in farmwork. Showing an entrepreneurial spirit, as a boy he had organized a group of neighboring kids in a business compact that involved exchanging mink skins for metal pins peddled by dealers bringing wares in wagons from nearby Atlanta. "Seems you can make anything from pins," Candler later remarked, recalling that speculating on their value had helped him save more than $100.[3] The secret to his future success was in recognizing such lessons about the value of servicing rural markets with city goods.

After the war, Candler went to live with his sister and worked for a while on a farm in Cartersville, Georgia, a small town some forty-three miles northwest of Atlanta, before finding a job at a pharmacy there in 1870. Candler had long been interested in becoming a doctor. In fact, he was named after a relative and one-time mentor, Asa W. Griggs, who had become a well-known physician in West Point, Georgia. But after two years on the job, he wrote to his namesake, "I think there is more money to be made as a druggist than as a physician," adding, "I know it can be done with a great deal less trouble of soul and body."[4]

The problem was Cartersville. "I want next year to go to a larger establishment in a larger place," Candler wrote Griggs. There he thought he could get "sufficient compensation for my services." It was off to Georgia's big city.[5]

Atlanta's population had grown considerably since the beginning of the Civil War. In 1860, the U.S. Census reported 9,554 people living in Atlanta, but ten years later, in the middle of Reconstruction, that figure had increased to 21,789 people. The migration of recently freed Black Americans played a significant role in this population boom. The city was still majority white in 1870, but the Black population stood at 45.6 percent of the total, up from 20.3 percent in 1860. This was a big change for Candler, who was coming from Cartersville, population circa 4,100, and whose hometown of Villa Rica only boasted 959 people (less than 10 percent of whom were Black).[6]

Candler found a job at the Atlanta drugstore of pharmacist George J. Howard, who would—somewhat reluctantly—become Candler's father-in-law when the young man married his daughter, Lucy, in 1878. Candler had a lot to manage. His father had died in 1873, so his mother, two brothers, and other family members were living with him and his wife in a house he rented on Pryor Street in downtown Atlanta (in the center of present-day Five Points). After the birth of their first child, Howard, in December 1878, the situation was getting cramped, so Asa moved to a bigger house a few blocks away.[7]

By all accounts, Candler was a relentless worker. He so impressed his boss that George Howard agreed to form a partnership with him in 1882. Howard and Candler, as the new drug firm on Peachtree Street in downtown Atlanta was called, sold patent medicines. Their inventory included "Cheney's Expectorant," a supposed cure-all for croup "whose formula," Candler admitted in an advertisement, is "unknown to me."[8] Other popular items included "Holmes' Sure Cure Mouth Wash and Dentifrice," billed as an "infallible cure for Ulcerated Sore Throat, Bleeding Gums, Sore Mouth and Ulcers." There was also "Botanic Blood Balm," which could supposedly do even more: "Have you sore throat, pimples, copper colored spots, old sores, ulcers, swellings, scrofula, itching skin, aches and pains in bones or joints, sore mouth, or falling hair? Then Botanic Blood Balm will heal every sore, stop the aches and make the blood pure and rich and give the rich glow of health to the skin."[9]

Candler's interest in these panaceas, like that of many other pharmacists in his industry, stemmed from personal problems. "You know how I suffer from headaches," he wrote to a relative in the 1880s. He was constantly in search of something that would help him deal with pain.[10]

And that is how he stumbled upon Coca-Cola, a temperance drink created in 1886 by another pharmacist, John Pemberton. Like Candler, Pemberton was attracted to the prospects of profits in the city. In the 1860s, he had established a respected pharmacy in Columbus, Georgia (population 7,401 in 1870), but in 1869 he had left in search of bigger business in Atlanta. When he arrived, Pemberton immediately ran into trouble, filing for bankruptcy in 1872 before suffering through two fires that consumed much of his inventory. He tinkered endlessly, creating new patent medicines, such as "Indian Queen Hair Dye" and "Globe Flower Cough Syrup," which he hoped would save him from ruin. But none of these concoctions really made Pemberton rich. Ultimately, he decided to try to imitate a drink called Vin Mariani that was becoming popular both at home and overseas.[11]

A Bordeaux wine infused with leaves from coca shrubs cultivated in South America, Vin Mariani was the brainchild of Angelo Mariani, a Corsican pharmacist, who first created his famous drink in 1863. Luminaries the world over, including Queen Victoria of England, President Ulysses S. Grant, and even Pope Leo XIII, were in love with Vin Mariani, writing of the drink's restorative properties in testimonials—thanks, at least in part, to its trace amounts of the cocaine alkaloid distilled from coca leaves.[12]

Pemberton's interest in coca and cocaine partially stemmed from his addiction to morphine. After being wounded while defending the city of Columbus against a Union incursion in April 1865—just days after Lee and Grant signed the terms of truce at Appomattox Courthouse—Pemberton started taking morphine to the point that one business associate labeled him a "drug fiend." Pemberton had read in medical journals about people who had supposedly overcome their dependency on morphine through coca consumption and was thus naturally intrigued by Vin Mariani. In 1885, he came up with a not-so-original knockoff called Pemberton's French Wine Coca and started selling it in Atlanta for one dollar a bottle.[13]

Pemberton's drink was not an exact copy of Vin Mariani, in part because Pemberton chose to infuse his drink with small amounts of crushed kola nuts. Grown and cultivated in West Africa for centuries, kola nuts contain caffeine, another stimulant Pemberton believed would help make his new patent medicine sell. And by all accounts it did. One advertisement in the *Atlanta Constitution* boasted that Pemberton was selling 500 bottles of his new product every day. It finally seemed he was going to make it after all.[14]

But right when things started to fall into place, Atlanta banned the sale of alcohol within city limits. Pemberton must have been stunned. He had finally found a formula that seemed to be a winner, but its alcohol content had suddenly become a liability. In the winter of 1885 and into the spring of 1886 he experimented with a temperance version that could work as a replacement, and thus was born Coca-Cola: a nonalcoholic drink containing both coca leaves from Peru and kola nuts from West Africa. He sold it as a syrup, rather than in bottle form, and asked soda fountain operators to provide the carbonated water. He advertised it as the "Ideal Brain Tonic."[15]

Asa Candler was intrigued. This "Wonderful Headache Specific," as one southern newspaper put it, seemed just the kind of thing he had been looking for. "Some days ago," Candler wrote to his son in the 1880s, "a friend suggested that I try [Coca Cola]. I did and was relieved." A few more samples had him convinced. "I determined to put money into it," he said, investing $500 in Pemberton's creation.[16]

Around this time, Pemberton fell ill and began looking to sell the rights to his new drink. Candler was just one of several investors who moved to purchase the "brain tonic" before Pemberton passed away in 1888. For months, Candler worked to gain full control of Coca-Cola, and finally he became the sole proprietor of the formula. In 1892, he officially incorporated the Coca-Cola Company.[17]

Business radiated out of Atlanta, with Coke syrup pulsing through railroad arteries coursing outward from the city. In 1890, 40 percent of the 8,885 gallons of syrup sold went to Atlanta soda fountains, but by 1891, Atlanta sales had dropped to 27 percent of the firm's total business, with most of the revenue coming from distant towns. The key to this growth was drummers, traveling salesmen who sought out business in soda fountains throughout the South and beyond. Some of these folks were seasonal cotton laborers from rural counties who were looking for work during lulls.[18]

In these years of expansion, Candler focused mainly on cities, but some folks in the Coca-Cola system began to think about the countryside as a profitable place to grow business. This was what Joe Biedenharn, a twenty-eight-year-old confectioner from Vicksburg, Mississippi, believed after his first few years selling Coca-Cola out of his candy shop. In 1890, he had bought a five-gallon keg of Coca-Cola syrup from Candler's nephew Sam Dobbs, and within a few years he was making a lot of money off the soft drink. "I saw the demand for Coca-Cola in town," he said, "and I thought it would be profitable in the country if I could only get it there." No one at the time was really bottling Coca-Cola for rural markets, in part because bottling infrastructure was still quite rudimentary. But Biedenharn had already been bottling water for various plantations on the outskirts of Vicksburg and figured he could do the same with Coca-Cola. Biedenharn became the first bottler of Coca-Cola, selling his first batch to the country market in 1894.[19]

Biedenharn used the natural highway of the Mississippi River to get his products to plantation communities. At that time, the only way to get goods to inland rural markets was by horse and wagon, and muddy roads often hindered draymen's ability to service the countryside. As a result, transporting syrup by boat became one of the fastest ways to spread Coca-Cola far and wide in the Mississippi Delta in those early years of Coca-Cola bottling.[20]

By all accounts, Candler was skeptical that bottling would bring in big business. When Biedenharn sent a case of his bottled Coke to the company boss, the Mississippi bottler received tepid approval to continue operations. In an 1896 financial report, Candler happily reported that Coca-Cola was "now sold . . . in almost all the cities of the United States, and in some of the cities of Canada, and in the city of Honolulu, H. I." He made no mention of rural markets. By 1898, Candler had built branch syrup factories in Dallas, Chicago, Philadelphia, and Los Angeles. At this point, his plants pumped out 214,000 gallons of syrup annually, roughly a hundred times 1889 output. The company seemed to be doing just fine without the bottling business.[21]

But if Candler was slow to recognize the value of servicing rural markets with bottles, others pushed country trade. In 1897, a local bottler of ginger ale, soda water, and cider in the small town of Valdosta, Georgia, over 200 miles south of Atlanta, became the second enterprise to bottle Coca-Cola for sale out in the country. Then, in 1899, two enterprising attorneys from Chattanooga, Benjamin F. Thomas and Joseph B. Whitehead, asked Candler for a contract to begin a large-scale Coca-Cola bottling enterprise. Thomas had first come to the idea while in Cuba during the Spanish-American War, where he saw local vendors hawking a bottled pineapple drink called Piña Frio. Candler was dismissive of the idea at first but ultimately gave Thomas and Whitehead the rights to distribute Coca-Cola in bottles throughout much of the United States.[22]

This decision changed everything. In just a few years, bottling plants popped up across the country. At first, franchises emerged in the big cities, Thomas and Whitehead signing contracts with businessmen that established franchises in Birmingham, Philadelphia, Louisville, and Kansas City. But in time the Coca-Cola bottling system spread to smaller towns, including places like Greenwood, Mississippi (population 5,836 in 1910), and Clarksville, Tennessee (population 8,548 in 1910). By the early 1910s, the company had even opened operations in Cuba, Jamaica, Bermuda,

Mexico, and the Philippines. All these bottling plants at first used mule- and horse-drawn wagons to truck bottles of Coca-Cola into the countryside. It was hard and arduous work, but it paid off. Putting up $3,500 to $5,000 to build a plant, some bottlers became multimillionaires peddling Coke.[23] By 1909, the company boasted 379 franchise bottlers operating throughout the country. These bottlers allowed Coca-Cola to access remote and isolated markets outside urban cores.[24]

The growth of bottled Coca-Cola caused problems for Candler, in part because his commerce now flowed across Jim Crow boundaries. At the turn of the century, Candler's drink still contained trace amounts of cocaine, a drug that white supremacists increasingly vilified as a stimulant contributing to Black crime in the American South. As historian Michael M. Cohen explained, "By selling the drink for the first time outside of the tightly segregated confines of the middle-class soda fountain, lower-class whites and African Americans could now get their hands on this potentially dangerous product. The consumption of Coca-Cola, and by proxy the consumption of cocaine, by these 'undesirables'—black men in particular—thus contributed to the next step in the cultural transformation of cocaine, this time from a necessary yet controllable medical tool to a threatening social menace." Bowing to the demands of the Jim Crow order, Candler decided to remove cocaine from Coca-Cola around 1903, leaving "decocainized coca leaf extract" as a flavoring component in his drink.[25]

Around this time, Candler also quietly removed references to coca leaves and kola nuts in company materials. In the 1890s, Candler had played up the connection to these agricultural commodities from the rural Andean highlands and the kola groves off Africa's Gulf of Guinea coast. But by 1905, Candler replaced such references with images of company factories churning out syrup across North America. People of color still harvested the leaves and nuts that gave Coca-Cola its distinctive flavor, but Candler no longer gestured to these rural worlds beyond the Jim Crow South—even if raw materials from these far-flung places were literally fused in the company's brand name.[26]

For years to come, Candler feared that the government would cry foul if the company ever fully removed coca leaves and kola nuts from the Coca-Cola formula. In a private letter to his son Howard written in 1908, Candler confessed the conundrum, noting that the secret formula "should contain enough . . . kola and coca to keep us from being charged by the government with being frauds."[27] As a result, these ingredients stayed in the drink, though the company refrained from referring to them in company advertising again.[28] In catering to the racial order of his day, Candler also cloaked his company's reliance on global commodities.

By the 1910s, the pace of Coca-Cola's conquest quickened with the arrival of motorized trucks. At first Candler had been a bit wary of this technology. In 1902, when his son Howard Candler, now in charge of Coca-Cola's New York office, wrote about purchasing a "locomobile" to speed up sales calls, Asa Candler wrote, "I very much fear that the use of a machine like that on the streets of New York City will ultimately get us into trouble" and "probably hurt some of you." "Laying myself liable to being charged with fogiism by the New York management," he explained in separate correspondence, "I never had any confidence in such machines. Like the Bycicle [*sic*] I have looked upon them as a fad only."[29]

But seven years later, Candler clearly had had a change of heart. At the opening ceremony for the 1909 Atlanta National Automobile Exposition, Candler quipped, "It was but yesterday when it was scarcely safe to operate an automobile in some parts of the rural sections of the country, so great was the prejudice against the horse-less carriage. This prejudice is rapidly passing, in fact is about gone." He spoke of how the car and truck would revolutionize every aspect of American commerce. "The man of the town," he said, "will find the refreshment of wholesome rural life 'far from the madding crowd's ignoble strife,' while the man of the country will find all the stimulation and advantages which contact with things urban can bring." In the end, he concluded, "The automobile will make the towns more countrified and the country more citified to the good of all concerned."[30]

The motorization of Coca-Cola's bottling system meant that salesmen could, as Candler put it, cover more "ground between suns."[31] The company in many ways became a pioneer in this new, swifter delivery method. In 1909, a company bottler in Knoxville purchased the first motorized delivery truck in the city, and on the other side of the Smoky Mountains, an Asheville bottler bought his first fleet of trucks in 1912 with an eye toward "making deliveries too distant to be served by the regular horse and wagon routes." The story was much the same in Palestine, Texas, where a bottler introduced the town to its first gasoline-powered delivery truck. This was a few years before Henry Ford's assembly line technology helped to bring down car and truck prices so that more middle-class Americans could afford them. In these early years, Coca-Cola bottlers spent considerable sums for vehicles that could serve what one Columbus, Ohio, distributor called "the country route." These new machines meant that Coke could now push soft drinks faster and farther beyond city cores, even if they had to travel down rough, rugged, poorly maintained roads. By the end of the decade, one bottler in Louisville, Kentucky, served a fifty-mile radius around his factory thanks to his fleet of trucks.[32]

Even if the strategy was not initially profitable, Howard Candler agreed with his father that they should continue to sustain "smaller branches" of their distribution network. They were, the younger Candler wrote, "a necessary evil so that our goods may be generally known throughout the country." With this mantra driving the business, by the start of World War I, Coca-Cola's annual sales had exploded to 6,767,822 gallons a year.[33]

As Coca-Cola expanded into the American countryside with its gasoline-powered trucks, the firm also began to change rural ecosystems in tropical lands overseas. The explosion in demand pushed communities that produced the drink's essential agricultural commodities to scale up production. Sugar was a case in point. By the end of the 1910s, Coca-Cola was the largest industrial consumer of sugar on the planet, using roughly 100 million pounds annually. Originally, the company sourced a great deal of this sugar from Cuban plantations, many managed by Milton Hershey of the Hershey Chocolate Company. In the early twentieth century, American sugar interests, led by Henry Havemeyer's U.S. Sugar Trust, denuded forests to make way for monocrop farms, now accessible via new railroads built on the Caribbean island by U.S. investors. While Coke was one of many American companies demanding sugar in the early 1900s, the company's enormous purchases helped fuel Havemeyer's and Hershey's rush to turn ecologically diverse ecosystems into monocrop farms.[34]

In the 1920s, when Coke began selling a sugarless concentrate to its foreign bottlers, the firm demanded that these overseas bottlers procure sugar locally. As a result, the ecological footprint associated with the firm's sugar purchases expanded. Brazilian bottlers sourced sugar from plantations in clearings made in Brazil's Atlantic forest, and European distributors procured their sweeteners from domestic beet growers. With its international expansion, Coca-Cola remained a product of the countryside, not only because its bottlers served rural markets back at home but also because its key ingredients came from farms all over the world.[35]

Candler recognized his firm's connection to worlds beyond the American South. His final years at Coca-Cola coincided with the completion of the Panama Canal, an event that Candler believed would forever change his native homeland. "Now that the Panama Canal has been opened," he said, "the Gulf of Mexico has become a great American lake, like the Mediterranean Sea has been for years a British lake, and in this American lake the South must always have a predominant influence." Looking to countries bordering the Pacific Ocean now accessible via the Panama Canal, he spoke of a "gulf-stream of commerce from the South" that "will soon be washing those distant and populous shores."[36]

Candler's home region, meanwhile, was primed to become a kind of global entrepôt. "We very properly seek to inform the world of our resources," he told Atlanta's chamber of commerce, "but we must be equally diligent in informing ourselves concerning the rest of the world. No city and no country can live to itself." He imagined his work as being like that of a "rain cloud, those 'merchants of the sky,' which gather moisture where it is too abundant and carry it where it is most needed." Atlanta sat in the middle of a commercial channel, he said, whose "genial current will quicken into life unimaginable and immeasurable industries."[37]

Candler still believed, like many New South boosters, that his home region would grow prosperous primarily by exploiting the Southland's "forests and fields and mines." During the Gilded Age and the beginning of the twentieth century, businessmen such as famed Atlanta newspaperman Henry Grady spoke loudly about how a "redeemed" South would rebuild from the ruins of the Civil War by establishing new industries that would take advantage of the Southland's natural resources. Candler echoed these claims. "The South," he rejoiced, "is becoming the manufacturer, as well as the grower of cotton." "Unto us," he told southern businessmen, "has been committed . . . a land where stones are iron, and out of whose hills we may dig brass and gems of priceless value."[38]

But in truth, Candler's Coca-Cola Company looked nothing like the textile factories, ironworks, and timber firms that became the darlings of prideful patrons of the New South creed. Coca-Cola was a new kind of business in the American South, a company whose value did not stem from resources mined from southern hills or grown in southern fields. Instead, Coca-Cola earned money by serving as a kind of capitalist conduit, coordinating the transport of resources cultivated and extracted from rural communities in faraway lands to consumer markets on the other side of the world. In other words, the company was a pioneer in worldwide logistics, long before the term became a buzzword in American commerce.[39]

And by playing the middleman in a global trade network that spanned the world, Candler made a lot of money. By 1914, at the age of sixty-three, he was a millionaire and one of the wealthiest men in the American South. His company was selling over 7 million gallons of syrup annually and bringing in roughly $9 million in revenue.[40]

By this point, Candler had grown weary of capitalist clawing. "I am close enough to the end to begin doing final things," he told his brother Warren in 1913. Devoted to the Christian faith instilled in his childhood, Candler felt uneasy about the riches he had accrued. "Life is far spent, and I have done next to nothing for my Lord. For the rest of the way I must do better." He told Warren that he didn't "want to have the money entrusted to my hands in a crumbling castle." He wanted to spend the rest of his years doing charitable works. In 1916, he handed the reins of his company to his son Howard and pursued public service through elected office, winning Atlanta's mayoral race in December 1916 and serving until 1919.[41]

Within three years, Howard sold his father's company to a financial syndicate run by Trust Company of Georgia president Ernest Woodruff. The price tag was $25 million. At the time, syrup sales continued to climb precipitously, and the future of business looked bright.[42]

But in the early 1920s, wild fluctuations in sugar prices hurt company earnings. And though the firm weathered this sugar crisis, by 1923 some believed the Coca-Cola Company needed new leadership.[43]

That is when thirty-three-year-old Robert Winship Woodruff arrived.

It is telling that the man who pulled Coca-Cola out of the doldrums of the early 1920s originally made his name in the trucking business. After all, the key to Coke's early success had hinged on motorized vehicles that allowed Coca-Cola to swiftly spread its syrup across vast territories of the country.

Born in 1889 in Columbus, Georgia, Woodruff had come to Atlanta with his father and family during the Gilded Age. Like Candler's, their move to Georgia's grand city was made in pursuit of bigger business and better opportunities. In 1904, Ernest Woodruff became the president of the Trust Company of Georgia, which afforded him the luxury of paying for his son Robert to attend Georgia Military Academy, a boarding school in College Park, Georgia. Woodruff was a bit of a wayward son who never really excelled at school. In 1908, he went off to Emory College in Oxford, Georgia, but dropped out before earning a degree. At the time, he knew his main goal, according to a close friend, was to "make a million dollars," but just how he would achieve this goal was not clear.[44]

In 1911 Robert went to work for the Atlantic Ice and Coal Company, a firm his father owned in Atlanta. He became a buyer for the company, which is how he met Walter White, the enterprising owner of White Motor Company based in Cleveland, Ohio. Being in the ice business in the hot climate of the American South, Woodruff became familiar with transport technology that allowed for swift shipment of perishable items. Woodruff felt that Atlantic Ice could increase efficiencies by replacing the firm's old horse- and mule-drawn wagons with White trucks, so he signed a contract with Walter White without telling his father. The move miffed the patriarch. Ernest and Robert had never really gotten along, and this act of insubordination made matters worse. Ernest decided he was not going to increase his son's pay, and in response Robert quit and connected with Walter White, who offered him a job on his sales team.[45]

Woodruff relished his new job. He started out as a salesman for the trucking firm in Atlanta but quickly rose to manage all the firm's sales in the Southeast. He became an advocate for improving roads in the rural countryside, meeting with county commissioners to promote the idea of equipping roadwork crews with motor vehicles.[46]

When the United States entered World War I, Woodruff joined the trucking division of the United States Ordnance Department, and there he convinced officials to buy hundreds of White Motor Company trucks. It was an early lesson in the profits businesses could earn through carefully crafted military contracts.[47]

Woodruff also sold trucks to the Coca-Cola Company and its bottlers. The pace of change was dramatic. In Los Angeles, for example, a Coca-Cola bottler went from having one motorized delivery truck in 1919 to having more than twenty-three in 1921. Not all Coca-Cola bottlers purchased White trucks, but many did, which helped Woodruff's sales record, drawing the attention of superiors. By 1922, he had risen to become vice president of White's firm, earning a healthy $85,000.[48]

Woodruff's ascension at the trucking firm coincided with a push by Progressive reformers to improve roadways in the United States, particularly in the American South. Beginning in 1915, a coalition of civic boosters, carmakers, public officials, and southern citizens came together to promote the development of the Dixie Highway, a circular road route that would connect the American South to communities as far away as the Great Lakes. This zeal for roadbuilding attracted both local and federal dollars to southern states well into the mid-1920s, and these funds financed the paving of roadways that Coca-Cola used to connect to serve its customers. To be sure, many southern roads would remain in disrepair for decades to come, especially once the Good Roads Movement of the Progressive Era began to wane with the onset of the Great Depression, but for a region that had been severely lacking in basic infrastructural development, these critical investments helped companies like Coca-Cola get their goods to markets far removed from city centers.[49]

While the construction of the Dixie Highway was underway in the 1920s, Ernest Woodruff and his allies at Coca-Cola became frustrated with Howard Candler's lack of vision and sought to replace him with Robert Woodruff. Apparently, despite the earlier tussles between father and son, Ernest still saw a place for his boy at the top of Coca-Cola. Accepting that position would require Robert to take a substantial pay cut and to leave a comfortable executive post at White Motor Company, but nevertheless, Woodruff decided to take the job at Coca-Cola. He said years later that he came back because he wanted to increase the value of his stocks. "The only reason I took the job," Woodruff reminisced, "was to get back the money I had invested." Some people who knew Coca-Cola's history well questioned whether this statement was true, because by 1923, the company seemed to be recovering from missteps in the early 1920s and the firm's stock price had climbed to its highest point in history. Nevertheless, Woodruff recalled that his plan was to make Coca-Cola stock rise even more, and then he'd "go back to selling cars and trucks."[50]

Woodruff brought a truck salesman's ethos to his job at Coca-Cola. Unlike Asa Candler, who only reluctantly agreed to bottle Coca-Cola in 1899, Woodruff believed wholeheartedly that transportable bottled drinks were the key to Coca-Cola's future. He wanted Coca-Cola to be "within arm's reach of desire," and he became obsessed with making sure that trucks carried the Coke brand to the most remote corners of the world. He focused a great deal of attention on gas stations, which numbered nearly 1.5 million in the United States by the end of the 1920s. No town was too small, no country road too narrow. It was said that in his final years running the firm, he would "squat next to a rural gas station's vending machine and count the number of bottle caps to see which percentage belonged to Coca-Cola."[51]

This emphasis on ubiquity paid off. In 1928, Coca-Cola sold more of its syrup to bottlers than to soda fountains. Coca-Cola bottling plants, totaling roughly 1,250 in 1929, were everywhere. There were now 600,000 retailers selling bottles, compared to 105,000 fountain dealers in the United States, Canada, and Cuba. The company boasted that it sold 9 million servings daily in these markets, with Coca-Cola salesmen traveling an estimated 2.2 million miles each year to service retail outlets. By the end of the Jazz Age, Woodruff wondered, "After national distribution, what?"[52]

The answer seemed clear: Coke's future lay in international sales. In the 1930s and 1940s, Woodruff took the system he developed in the United States—one focused on creating access to bottled Coca-Cola in remote markets—overseas, launching the Coca-Cola Export Corporation in 1930, which aggressively courted foreign businessmen interested in becoming Coke distributors. The big breakthrough came in World War II, when Coca-Cola secured special contracts with the United States military to service American GIs overseas.[53]

During the war, the company sold an estimated 10 billion servings of Coca-Cola to American soldiers, some of it bottled at sixty-four new overseas bottling plants that had been built with the assistance of the War Production Board and the quartermaster general's office.[54]

Coke transacted a good portion of its military sales in the American South, in large part because the federal government invested millions of dollars in the region during the war to build army barracks, navy bases, and other military installations. In the 1940s, the U.S. military ultimately sited over 65 percent of all navy and army posts in towns and rural communities throughout the American South. The remoteness of many of these bases was no problem for Coca-Cola, a firm that had for decades mastered the art of getting its beverages to small communities throughout its home region. After all, even at the start of World War II, roughly two-thirds of all southerners worked on farmland or in towns of less than 2,500 people. Coca-Cola had already brought Coke to most of these rural communities, charting paths to profits down the most rural of country roads.[55]

As military personnel flooded into the South, so did cash that could be used to buy nonessential goods like Coca-Cola. Servicemen had disposable income to spend on soft drinks, and so did industrial workers in the new defense and munitions factories that sprouted up in southern towns. By this time, wealthy landowners had used New Deal subsidies to invest in labor-saving machinery and evicted tenants from farmland, which caused rural families to flood into southern towns to take advantage of the new defense jobs that had come to the region. White Americans secured most of this factory work in the South, in large part because the federal government did little to ensure that Black citizens received equal treatment from employers; but for those fortunate enough to get defense jobs, the pay was well above the going rate in the American South. For many families benefiting from this infusion of federal money, five cents was now a small price to pay for a little refreshment.[56]

With the U.S. military providing money, transport, and personnel, Coca-Cola also spread internationally to places where there was very little infrastructure to support it. Officers at overseas military bases who requested Coca-Cola constantly spoke of their isolation from industrial centers. They demanded bottle washers and filling equipment, carbonation tanks and conveyer belts. A surgeon stationed at a base in Trinidad spoke of the need for chemicals to treat the water "in order to produce Coca-Cola which will meet the specification set up by both the Coca-Cola Company and the United States Army Medical Corporation." Others demanded "refrigeration equipment" and coolers. A "Coca-Cola concessionaire" in Paramaribo, Dutch Guiana (today the Republic of Suriname), wrote to a quartermaster requesting a "three-quarter ton, panel Chevrolet truck selling for approximately $832.35" as well as a large "chilling compartment" to keep drinks cold. He made these requests because he said it was impossible to secure "delivery on the truck and cooling equipment locally."[57]

Coca-Cola's ecological footprint grew immense. The company was bringing more than just its drinks to faraway corners of the world; it was stimulating investment in a host of industrial infrastructure—trucks, refrigerators, pumps, and pipes—that allowed it to turn local natural resources—water from aquifers, sugar from local farms—into its ice-cold beverages. The U.S. government helped pay for much of this growth during World War II, and so did independently owned foreign bottlers enticed into the Coke system by the prospects of profits. The key for Coke was always staying at arm's length from the actual business of making the finished products it hoped would be only an "arm's length from desire." Local businessmen overseas, just like domestic bottlers at home, took on the risks and the costs of getting bottling plants up and running, investing capital lent from local banks to pay for all the machinery that would ensure swift delivery of finished beverages to customers.[58]

By 1948, there were over 354 international bottlers operating around the globe. Two years later, *Time* magazine spoke of Coca-Cola's "near-conquest of the world," noting how the firm's beverage had transformed rural communities across the world, bringing "refrigeration to sweltering one-ox towns without plumbing." What *Time* understood was that Coke was more than just a beverage company. The firm stimulated purchases of a host of equipment and machinery that made it possible to serve its ice-cold drinks even in the most remote parts of the world. In time, that infrastructure became the biggest part of the firm's global greenhouse gas emissions.[59]

But fears about climate change were not yet a concern in the United States or any other part of the world. For now, *Time* dubbed Coke's growth an unassailable success story. "It is not a product of the vast natural resources of the land," the magazine marveled, "but of the American genius for business organization."[60]

Chapter Two

WITHIN ARM'S REACH OF DESIRE

It was fall 2009, and Jeff Seabright, vice president of water and environment at Coca-Cola, was worried. He had just walked up to the top floor of Coca-Cola headquarters on North Avenue in Atlanta, Georgia, and he was standing outside the executive suite of CEO and chairman Muhtar Kent, where he planned to deliver some bad news: a Japanese manufacturer that had assured Coke it could make refrigerator systems that used CO_2 as a compressor gas had just reneged on its promise to the company.[1]

Seabright, now in his fifties, was no stranger to high-pressure situations. He had been the chief sustainability officer at Coca-Cola since 2003, when the company was reeling from a series of highly publicized battles with Indian villagers in Plachimada, Kerala, who claimed that a Coke bottler was threatening their water supply. Before that he had been the vice president of policy planning at Texaco, where he had tried to steer that firm toward a more climate-friendly energy future. And in the late 1990s he had served as the executive director of the White House Task Force on Climate Change during the bitter debates over the Kyoto Protocol. In short, Seabright was a seasoned negotiator, but that did not make him any less nervous in this moment. He was, in his own words, "frightened."[2]

As CEO, Kent was just days away from making a major announcement in New York with recently retired Greenpeace executive director Gerd Leipold. Kent planned to reveal that Coca-Cola would be buying hundreds of thousands of new CO_2-filled refrigeration units that would replace the standard hydrofluorocarbon (HFC) coolers the company had been using for decades. HFC refrigerators had come under fire in recent years because the chemicals many of these coolers contained had a global warming potential (GWP) that was over a thousand times higher than CO_2. The problem was that CO_2 refrigeration systems operate at extremely high pressures and are therefore more difficult to manufacture and maintain than equivalent HFC technology. Seabright had been thrilled to find an Asian supplier that would partner with Coca-Cola to do the hard work of creating a line of CO_2 coolers, but without the Japanese partner, it now seemed all was lost.[3]

Seabright and Rick Frazier, the senior executive in charge of Coca-Cola's supply chain, passed through the door to Kent's office but made it no further. The executive did not ask them to sit as Seabright explained the situation about the failed Japan deal.[4]

Moments later, Kent—a man well-known for his "autocratic leadership style"—began pounding his fist on the table. "I made my commitment to Gerd that I will do this," he yelled to the two men standing in the entryway. "I am going to be at that event with him next week in New York. We are going to make this announcement. You are going to figure out how we get from here to there. That is your job. Now get out!" He then pointed to the door.[5]

Seabright remembers being dazed. The thought probably crossed his mind that he was on the verge of losing his job. He headed back down to Frazier's office on the twenty-fifth floor of the Coca-Cola tower, and the two businessmen just stared out at the Georgia Tech campus below for several minutes in total silence. What would they do now? They had no idea how they were going to fix this.[6]

The next couple of days were a flurry of activity: Seabright and Frazier began a series of "chaotic" conversations with potential CO_2 compressor suppliers. They knew they had to land a deal. Muhtar Kent had told them to spare no expense. And in the end, the frenzied hustling worked. Seabright was able to find a company in China that would take Coke's contract. It was an extremely stressful week, but Seabright got to keep his job. And Muhtar Kent got to make his big announcement with Greenpeace.[7]

It might seem odd that an issue so seemingly simple as refrigeration would create so much anxiety in the highest offices of Coca-Cola. But the truth is, Kent and Seabright both knew that refrigeration modification was central to Coca-Cola's efforts to enhance its environmental profile. After all, when Kent made his pledge to Greenpeace in 2009, refrigeration had become the single largest source of greenhouse gas emissions for the company. If Coke was going to go green, it was going to have to do something about all its chilling units spread across the world.[8]

To understand why refrigeration became such an outsized component of Coca-Cola's greenhouse gas footprint and why Coca-Cola initially chose CO_2 refrigerants for its new coolers requires heading back to rural roads. In its quest for ubiquity, Coca-Cola helped create a vast network of cooling equipment that in time would play a significant role in warming our planet.

The idea that Coca-Cola should be sold "ice-cold" was not always a given. In fact, back in the late 1890s and early 1900s, some soda fountain operators advertised "hot Coca-Cola" to patrons to increase sales in the winter months. Nevertheless, most soda fountain clientele in the early 1900s expected a chilled beverage when they ordered a Coca-Cola, especially in the hot American South, and this had important implications for Coca-Cola's salespeople, who had a harder time selling syrup in frosty weather. In 1922 the company—still frustrated with the common consumer perception that Coke was a summer beverage—developed the slogan "Thirst Knows No Season," hoping to convince customers that December was just as good a time to drink cold Coke as August. Eventually, the sales strategy worked. Summer remained a peak season for Coke sales, the firm profiting from warm climatic conditions that drove consumers to seek refreshment for decades to come, but the drink also became increasingly popular in the chilly, dark days of the holiday season.[9]

The fact that customers came to see "ice-cold" Coca-Cola as the natural order of things, whether in summer or winter, had a lot to do with the physics involved in making soft drinks. In the early decades of growth, Coca-Cola bottlers invested in multiton mechanical refrigeration units capable of cooling the carbonated water put into bottles. The goal was to keep the temperature below 40 degrees Fahrenheit so that the carbonation would remain in the sealed container. Bottlers also sought to keep their beverages cold at the point of sale, not only because they believed it made Coca-Cola taste better but also because, as one distributor put it, "the low temperature of the drink locks the carbonation in," ensuring that the beverage has its signature fizz when popped open. In the American South, where summer temperatures could rise well above 90 degrees Fahrenheit, finding appropriate cooling equipment to keep Coke cold was particularly important.[10]

Fountain operators, not just bottlers, worked to make sure that Coca-Cola was served at optimal temperatures (between 34 and 40 degrees Fahrenheit) in part to preserve the physical properties of the drink. In fact, in the latter half of the twentieth century, McDonald's became obsessive about the temperature at which it sold its Coca-Cola, chilling not only the carbonated water that came out of its fountains but also the syrup that was mixed with that water. Today, if you get a Coca-Cola from McDonald's, it is served colder (around 34 degrees Fahrenheit) than beverages sold at other fountain outlets. When a Coca-Cola bottler heads out on a service call to McDonald's or other fountain vendors to address customer complaints about "flat" beverages lacking carbonation, one of the first things a serviceman does is check to make sure the temperature of the carbonated water is optimal.[11]

Refrigeration equipment today is relatively inexpensive compared to the investments Coca-Cola bottlers and retailers had to make in the early years of Coke's growth. Back then, in the 1910s and 1920s, mechanical refrigeration was just being perfected by American manufacturers. The first mechanical ice plants emerged in the United States in the 1860s just after the Civil War, and the brewing industry became the primary customer of this machinery in the 1870s and 1880s (again due to the demands of trapping carbonation in their brews). The American South was a leader in the development of mechanical refrigeration, in part because the region's access to natural ice supplies harvested from New England ponds had been cut off during the Civil War. But by 1900, many companies across the country in need of refrigeration began switching away from natural ice to mechanical technologies—electrically powered compressors and evaporators—to meet their cooling needs.[12]

Coca-Cola bottlers spent considerable capital on refrigeration in the early years, buying two-to-ten-ton mechanical refrigeration units to cool the water that went into their bottled beverages. Prior to the late 1930s, these huge units—described in mesmerizing detail by visitors to bottling facilities, especially in small towns—used machines filled with refrigerants common for the day, such as sulfur dioxide, methyl chloride, and ammonia. Many of these chemicals were toxic to humans in the concentrations they were used; sulfur dioxide was one of the preferred refrigerants in part because the chemical smelled so bad, alerting users to potentially dangerous leaks. In short, these were mammoth machines filled with chemicals that required substantial investment, both for the initial purchase and for the continued maintenance and upkeep thereafter.[13]

These big refrigeration units were cost prohibitive for most of the mom-and-pop shops where Coca-Cola sold its beverages in the early decades of the twentieth century. As a result, well into the late 1930s, vendors used ice in relatively rudimentary coolers to chill Coca-Cola. Through the Great Depression, most consumers also used ice to cool Coca-Cola in their homes, in part because advancements in refrigeration technology had not made mechanical units small enough or cheap enough for many families to buy.[14]

But the refrigeration industry changed dramatically in the 1930s when major manufacturers—Kelvinator, Frigidaire, General Electric, and Westinghouse—improved assembly-line manufacturing methods that helped bring down the price of smaller refrigeration units. New Deal programs, such as the Rural Electrification Administration and the Tennessee Valley Authority, also gave the industry a boost, creating new markets for electrical appliances in rural markets, especially in the American South. By the end of the 1930s, the price of a household refrigerator had dropped from an average of about $600 in 1920 to $152 in 1940.[15]

Many of these new refrigerators used new chemicals developed by Thomas Midgley Jr., a chemical engineer working for Dayton Engineering Laboratories (Delco), a research group owned by General Motors and based in Dayton, Ohio. In 1930, Midgley developed chlorofluorocarbons (CFCs), a new, nontoxic, nonflammable refrigerant that was stable and did not degrade rapidly over time. Frigidaire, also owned by GM, soon created a partnership with DuPont to begin manufacturing this new chemical for commercial markets. These CFCs, sold under the brand name Freon, were initially expensive to produce, but DuPont sold them below cost to spur consumer demand. Over the course of the next couple of decades, Freon gradually replaced most of the older refrigerants used in cooling equipment, largely because these CFCs were seen as safe and reliable.[16]

Coca-Cola was a big buyer in this new era of CFC refrigeration. By the start of World War II, company bottlers—of which there were now well over 1,000 across the country—were already using CFC-filled condensers to chill the water going into company beverages. In addition, vending machine firms, such as Vendo based out of Kansas City, Missouri, had also begun developing electrical machines for Coca-Cola (though many of Vendo's coolers were still chilled with ice through the Great Depression). By the 1940s, Coca-Cola also contracted with companies such as Westinghouse to develop Freon-filled refrigerators for the consumer market that could be put in service stations, corner shops, and even factory floors across the country.[17]

The war temporarily put a halt to civilian sales of CFC equipment, but Coca-Cola continued to find ways to expand sales of refrigerated beverages in the United States in part by serving many of the military installations that the federal government built in the American South during the war. Rationing restrictions did not apply at these locations, which meant that if Coca-Cola was serving troops, whether at military bases at home or abroad, they could also request refrigeration equipment that otherwise would have been unavailable to the firm due to government rationing rules. In short, military investments, especially in the American South, helped keep Coca-Cola cold in the United States from 1941 to 1945.

And when federal rationing programs for civilian markets expired at the end of the war, companies such as Vendo and Westinghouse dramatically increased sales of Freon-filled refrigerators and vending machines to Coca-Cola bottlers and retailers.[18]

Robert Woodruff, the president of Coca-Cola, now into his second decade at the helm of the firm, must have been pleased. Having long promoted the mantra that Coca-Cola should be within an "arm's reach of desire," Woodruff wanted Coke customers to be able to pop open a chilled bottle wherever they were and whatever the time of day. Now, thanks to New Deal rural electrification programs and affordable mechanical refrigeration technology, Woodruff was closer to achieving his dream.[19]

And not just at home. As we have seen, the military contracts Coca-Cola signed with the U.S. government in the 1940s included provisions for transporting chilling equipment to Europe, North Africa, and the Pacific theater. By 1948, there were 354 international Coca-Cola bottlers and 1,056 U.S. distributors, all spreading refrigeration units into remote markets.[20]

Over the course of the next two decades, CFC-filled Coke coolers and vending machines, became increasingly critical to Coca-Cola's growth, in part because the firm began diversifying into new perishable product lines. Take Minute Maid, the orange juice and Hi-C fruit drink maker, which Coca-Cola acquired in 1960. Unlike soft drinks, some of Coke's fruit drinks needed to be kept cool in order not to go bad. Losing carbonation and a "zippy" taste was one thing; spoilage was an entirely different matter. In the decades ahead, refrigeration became critical to a new Coke company no longer wedded to one soft drink brand.

But in time, it became clear that there was a serious ecological problem with Coca-Cola's dependence on widely dispersed CFC chilling infrastructure. In 1974, UC Irvine scientists F. Sherwood Rowland and Mario Molina produced research showing that CFC refrigerants were depleting the ozone layer. DuPont, the largest producer of CFCs in the world, protested, fearing regulation that would destroy a profitable business valued at $8 billion in 1975, but by the 1980s, the U.S. EPA had already moved to ban CFCs in aerosol cans (the chemical was also used as a propellant), and there was momentum toward banning CFCs from refrigeration equipment as well. In 1987, world leaders met in Montreal to sign a protocol that banned the further use of CFC refrigerants, and in 1988, President Ronald Reagan signed the agreement.[21]

DuPont promised it had a solution to the "hole" in the ozone layer: new refrigerants called hydrofluorocarbons, or HFCs. These chemicals, which DuPont stood to make lots of money from, did not contain the ozone-depleting chlorine atoms found in CFCs and had the added advantage of easily replacing chlorofluorocarbons in existing refrigeration equipment.[22]

But there was a problem. HFCs, while lacking the ozone-depleting properties of CFCs, nevertheless had exceptionally high GWP; some varieties of HFCs had more than 3,400 times the power to affect planetary warming than carbon dioxide did. With climate change becoming an ever more important issue in the 1990s, organizations such as Greenpeace railed against companies that began making the switch to HFCs, arguing that the new compounds might be good for the ozone layer but were not good for a warming planet.[23]

And the organization did more than just protest. In the early 1990s, Greenpeace began working with manufacturers to develop what became known as "Greenfreeze" refrigerators that used propane, isobutane, and other hydrocarbons with low GWP ratings—over 1,000 times lower than many HFCs. Greenpeace dubbed hydrocarbons "natural" refrigerants because they were not man-made like DuPont's HFCs. First sold in Germany in 1993, "Greenfreeze" technology quickly became the target of chemical industry attacks, with companies invested in HFC technology charging that these new refrigerators would be "potential bombs" that could explode due to the flammable compounds they contained. Nevertheless, several German refrigeration manufacturers adopted the Greenfreeze technology, and units began to appear across Europe.[24]

But back in Atlanta, Coca-Cola managers were committing the firm to HFCs in large part because of Coke's long tradition of serving countryside communities. The firm worried that using hydrocarbons in these remote markets could be extremely dangerous because there would be less technical oversight to prevent explosions. In short, Coke's business model, which depended on deep penetration into rural communities, shaped the firm's decision about which refrigerant to use in the years ahead. It was going with HFCs for the foreseeable future. And the transition was going to take a lot of work. In 1991, the company said it had more than 2 million vending machines out in the world, and over 6 million pieces of other sales equipment, including millions of coolers and refrigerators. The work of converting all those units away from CFCs began in earnest in 1992.[25]

Coca-Cola engineer Bryan Jacobs remembers that time well. He had just graduated from Georgia Tech and was looking for job opportunities that would allow him to stay in the Peach State because he planned to wrestle in the 1996 Olympics in Atlanta. In 1992, Jacobs had worked for Anheuser-Busch in Cartersville, Georgia, part of an Olympics job program that offered U.S. athletes flexible scheduling options as they trained for that year's summer games in Barcelona. He sought and received the same opportunity at Coca-Cola a year later and began working in the company's newly created environmental technical affairs department.[26]

Coca-Cola had just announced that it was committed to two major environmental initiatives: installing wastewater treatment technology at all its production facilities around the world and phasing out CFCs. Jacobs worked on both assignments but became a particularly important player in the refrigerant phaseout program.[27]

The first month he was on the job, Jacobs received a call from an engineer at a Coca-Cola design facility in Essen, Germany, who told Jacobs the company should be looking at hydrocarbons, not HFCs, as it transitioned to new refrigerants. The German engineer had learned about the new "Greenfreeze" technology because Greenpeace had been pushing it so heavily in his home country. If Coke decided to choose hydrocarbons over HFCs, the German engineer said, the firm could radically reduce its greenhouse gas footprint.[28]

But Jacobs said Coca-Cola engineers in Atlanta were "pretty dismissive" at the time because research conducted by the automotive industry on HFCs had shown that these man-made chemicals were nontoxic, nonflammable, and odorless. These properties were particularly important for a firm that placed its refrigeration equipment in remote places around the world. Jacobs explained that Coca-Cola was reluctant to put refrigeration units containing flammable compounds in rural communities that lacked a "highly developed service infrastructure that can be adequately trained to safely handle a flammable refrigerant." Coke's success at practicing country capitalism, in other words, was a liability when it came to dealing with its refrigeration emissions problem.[29]

Throughout the 1990s, Greenpeace continued to promote hydrocarbons over HFCs, even hosting the "Ozone Safe Cooling Conference" in Washington, D.C., which brought manufacturers together with environmentalists to discuss alternatives to HFCs. In Europe, Greenpeace was able to get German firm Bosch to sell hydrocarbon-filled refrigerators by the mid-1990s. By 2001, essentially all new refrigerators manufactured in Germany used hydrocarbon refrigerants. Asia also saw growth in Greenfreeze technology adoption when Haier, the home appliance company based in Qingdao, China, agreed to switch to non-HFC refrigerants. Global momentum continued into 1997, when Kyoto Protocol signatory countries agreed to begin phasing out HFCs.[30]

All the while, Coca-Cola was still transitioning millions of old CFC-filled coolers to new HFC units. By 1997, the company reported that there were over 6.6 million vending machines, refrigerators, and dispensing units in Coca-Cola's system. And even though Coke had over 180,000 gas-guzzling trucks in its bottling network, it was cooling equipment, not diesel vehicles, that represented the largest share of the firm's greenhouse gas emissions.[31]

For years, Greenpeace had seen chemical companies, such as America's DuPont and Britain's Imperial Chemical Industries, as prime targets in its anti-HFC campaigns, but by the end of the 1990s, activists decided to switch tactics. They focused on the upcoming 2000 Olympics in Sydney, Australia, and began organizing a mass demonstration against big corporate sponsors like Coca-Cola and McDonald's that planned to use HFC refrigerators at the games.[32]

Corin Mallais, a longtime Greenpeace activist from Great Britain, was one of the point people on this campaign. He recalled that Greenpeace had planned some bold actions for Sydney, including having activists jump onto container ships to try and stop the delivery of HFC refrigerators to Australia. That spectacle never happened, but Mallais and others produced powerful propaganda for the international event, including posters that depicted somber polar bears drifting out to sea on melting blocks of Arctic ice below the headline "Enjoy Climate Change." A caption below the image read, "Coca-Cola's use of HFCs to cool its drinks contributes to climate change. Ban HFCs." Other protestors slapped stickers on Coke vending machines that highlighted Coke's reliance on climate damaging refrigerants. And Greenpeace published a pamphlet called *Green Olympics, Dirty Sponsors* that exposed the tremendous impact HFC refrigerators had on climate change while simultaneously launching a website called Cokespotlight.org—one of the organization's first internet-based campaigns—that allowed visitors to send protest emails about this issue directly to Coke chairman and CEO Douglas Daft.[33]

The fact that Daft was now at the top of Coca-Cola mattered. He was from Australia and was particularly upset about the thought of his company being embarrassed in his home country. "It got attention," said Bryan Jacobs, speaking of Greenpeace's actions. That summer, just months before the games were due to commence, Daft met with Greenpeace and made a bold commitment that by the next Olympics games in Athens, the firm would no longer buy HFC refrigerants for new cooling equipment. It also rushed out a few dozen HFC-free refrigerators for the Sydney games, but this was seen as a small gesture given the fact that company had over 1,800 refrigeration units at the 2000 Olympics.[34]

Jacobs—who was by the time of the Sydney games part of Coke's newly formed eKOfreshment team, created in 2000 to explore environmental sustainability objectives and so named because of Coke's stock ticker abbreviation—said his firm's pledge was too ambitious. There was simply no way for Coca-Cola to meet its target in four short years. Jacobs—who was used to thinking in quadrennial timeframes, having trained for the Olympics himself—had worked hard to get Coca-Cola prepared for the games in Greece, but in the end, the company was only able to deploy roughly 500 HFC-free units in Athens, far short of its 100 percent goal.[35]

One reason Coca-Cola was slow in getting to its target was that it had selected an unconventional refrigerant for its coolers: CO_2. This choice may seem surprising given that CO_2 is a major greenhouse gas contributing to climate change. But like hydrocarbons, CO_2 had a global warming potential that was over 1,000 times lower than many HFCs. More importantly, it was nontoxic and nonflammable. Coca-Cola felt it had to invest in CO_2 technology in part because it needed a safe alternative to hydrocarbons, which could potentially explode in remote and rural sales locations. The company did a study in the 2000s that showed that if it chose to use propane or isobutane in rural markets, the firm would have at least "one catastrophic event per year." This was simply not an option. The company had to get refrigeration companies to invest in CO_2 technology, and that was not going to be easy. CO_2 systems were expensive to make, in part because they had to be built to operate at higher pressures.[36]

Whether manufacturers—such as Embraco, based in Brazil, or Imbera, based in Mexico—would take on this task in 2000 was still unclear. The market for CO_2 refrigeration was basically nonexistent back in the early 2000s. In fact, when Coca-Cola announced it was going HFC-free, the company took heat from members of the Air-Conditioning and Refrigeration Institute (ARI)—which included compressor companies and refrigerant makers such as DuPont and Honeywell—who assumed Coca-Cola would be turning to isobutane or propane. Jacobs said that at the 2000 ARI conference in Washington, D.C., firms "took turns berating us, accusing us of not knowing what we were doing," and saying, "You're going to put these flammable, explosive refrigerants out there, somebody is going to get killed!" But Coca-Cola quickly explained that it was exploring other alternatives to hydrocarbons and was looking for partners to help them build what they needed.[37]

And to show potential collaborators they were serious, in 2004 Coca-Cola united with Unilever and McDonald's to create Refrigerants Naturally!, a new industry group committed to converting refrigerators to HFC-free technology. This was important because it meant that manufacturers now knew there would be big buyers willing to purchase HFC-free refrigerants if they decided to invest in new technology.[38]

Greenpeace worked with Coca-Cola during these years to achieve its objectives, drawing the ire of some within the organization who believed Greenpeace was "selling out" by working with big business. Activist Amy Larkin saw it differently. During the early 2000s, she was the director of Greenpeace Solutions, a newly created unit of the international environmental organization tasked with forming partnerships between Greenpeace and large multinationals to solve environmental problems. When she first met with Coca-Cola's refrigeration team, she expressed her willingness to work with the company but also made it clear that she understood why she even had a place at the negotiating table. As she often said to businesses leaders she met with, "The only reason I'm here is because my colleagues jumped on your roof." Bryan Jacobs remembers a slightly different statement from a Greenpeace organizer who apparently told Coke leadership that his organization had "agreed to dance with the Coca-Cola Company on Refrigerants Naturally! . . . but we reserve the right to dance on [Coca-Cola] on other subjects."[39]

One place where Greenpeace proved helpful to Coca-Cola was in negotiations with the U.S. EPA. When Coca-Cola and its partners first launched Refrigerants Naturally!, the EPA had not approved isobutane, propane, or CO_2 for use in refrigeration technology sold in the United States. Larkin remembers going in to "beat up" on EPA officials hesitant to adjust refrigerant regulations (in part due to concern about the hazards of hydrocarbon refrigerants). It took some time for the agency to act, but in 2009 the EPA officially registered CO_2 as an approved refrigerant for use in U.S. markets, and in 2011 it added propane to the list. The EPA ultimately acknowledged that Coca-Cola's steadfast lobbying played a significant role in shaping the agency's decisions.[40]

By this point, Jeff Seabright was vice president of environment and water at Coca-Cola and was working with Bryan Jacobs and others to execute the HFC phaseout. He had collaborated closely with Greenpeace and was impressed with how collegial negotiations had been. He recalled that at one point Coca-Cola had made an HFC-elimination target, but it soon found that the cost of achieving its goal by the date it had pledged was going to be "prohibitively expensive." So he decided to fly to New York City to meet with Amy Larkin and her Greenpeace team, and he asked them to sign a nondisclosure agreement so he could show them Coca-Cola's confidential spreadsheets detailing the company's financial position and the trouble Coca-Cola was having securing enough HFC-free refrigerators at a price point that made sense for the firm. Grudgingly, Larkin and her associates signed the NDA, and Seabright made his case. "I just opened the books," he said, "and showed them where we were and the fact that we were committed but we're not stupid." Seabright was struck by their response: "They were like, 'Oh, we get it. You're in business and you're on it, you're committed, we understand. . . . We're not happy with where we are, but we're not going to raise a critical voice.'"[41]

Seabright was relieved. He, like many others within Coca-Cola, respected the power of Greenpeace. "There's no question," he said, "without pressure from Greenpeace, nothing would have happened." He remembered people at Coca-Cola saying that Greenpeace was going to "eat us for lunch" if it didn't honor its HFC-free pledges.[42]

Which is why Muhtar Kent, Coca-Cola's CEO, was so keen on honoring his commitment to the international environmental organization when Seabright came to his door in 2009 with the bad news that Japanese manufacturers were not going to be able to deliver enough CO_2 refrigeration systems to meet company pledges. He knew that the company had to meet its pledge to Greenpeace or face scathing ridicule that would do real damage to the company's image. Refrigeration was simply too big an issue for the company to ignore.

And Kent did more than just urge his employees to commit to an HFC-free future. He got other big businesses to do the same. At a 2010 conference in Cancun, Kent led the Consumer Goods Forum, a consortium of some 400 consumer goods companies, to adopt a resolution to eliminate HFCs from all new refrigeration equipment by 2015. Amy Larkin spoke at this event at the invitation of Kent, and she remembers how big a deal it was. Now hundreds of companies—not just Coca-Cola, McDonald's, and Unilever—were pledging to end their dependence on HFC gases.[43]

Coca-Cola's decision to push for natural refrigerants had real effects on the cooling equipment industry. In 2014, Coca-Cola happily announced that it had introduced its 1 millionth HFC-free cooler into the environment. Two years later Refrigerants Naturally!—which now included PepsiCo and Red Bull in addition to original members Coca-Cola and Unilever—boasted that they had collectively put 5.5 million HFC-free coolers into global markets. Refrigerants Naturally! also supported the 2016 Kigali Amendment to the Montreal Protocol, which called for an 80–85 percent reduction in HFC use by the end of the 2040s. All these efforts made natural refrigerants more viable for other companies, including automobile manufacturers, which began slowly switching away from HFCs in motor vehicle air-conditioning systems in the late 2010s.[44]

Yet there was still a long way to go. In 2014, only 30 percent of all Coca-Cola chilling equipment in global markets was HFC-free. And in 2015, Coca-Cola missed its procurement target of using only natural refrigerants in its cooling systems. As late as 2017, only 65 percent of new coolers purchased by Coca-Cola contained natural refrigerants, and in 2019, 18 percent of its new cooler purchases contained HFCs.[45]

Still, Coca-Cola's sustainability team—folks like Seabright and Jacobs—had clearly tried hard to move the soft drink giant away from HFCs. And work continued after they were gone. By the end of the 2010s, Coca-Cola was able to speed up its conversion to natural refrigerants by loosening its preference for CO_2 condensers and buying more and more propane and isobutane units. This decision was driven by new evidence that hydrocarbon-based coolers could be operated safely in many markets, including more rural ones. Coca-Cola was making a good-faith effort to get rid of as many HFC units as it possibly could.[46]

Yet the question remained: Was Coca-Cola really on the path to a sustainable future? The numbers should have forced the company to pause. In 2018, Coca-Cola still reported that refrigeration represented its single largest contribution to global warming, with refrigeration-related emissions accounting for more than a third of its total greenhouse gas footprint. Coca-Cola said it was dealing with the problem, not simply by committing to natural refrigerants but also by installing new "intelligent energy management devices" that saved its customers more than $400 million a year in electricity costs and reduced greenhouse emissions by 3.1 million metric tons of CO_2 equivalent a year.[47]

Still, it seemed that all those coolers, no matter how energy efficient they were and no matter what refrigerants went into them, still contributed to a tremendous proportion of Coca-Cola's overall emissions. Given this reality, one wonders whether Coca-Cola's sustainability team ever considered the option of eliminating some of its refrigeration infrastructure altogether.

Jeff Seabright fielded this question in 2021. He was no longer at Coca-Cola, having left the firm in 2014. At the time, Seabright said he was ready for a change. He candidly admitted that he "was having an increasingly difficult time with the obesity/health issues and the lack of ability of the company to take it seriously." He felt company executives were holding an "indefensible" position when it came to Coca-Cola's contributions to health problems, one he simply could not support. So he became the chief sustainability officer at Unilever before founding an organization called IMAGINE, which sought to "develop and accelerate positive change in our global systems."[48]

Seabright, who clearly cared deeply about environmental issues, said the idea that Coca-Cola would ever consider getting rid of some of its Coca-Cola cooling systems was absurd. "No," he said emphatically, Coca-Cola never, ever considered reducing its greenhouse gas footprint by reducing the availability of Coca-Cola coolers in metropolitan or rural markets around the world. He invoked the words of Coca-Cola boss Robert W. Woodruff, saying, "The business model, 'within arm's reach of desire,' immediate consumption. . . . That's all part of kind of the magic of why Coke continues to be [able to] sell sugar water after a hundred and thirty years." Executives wanted "more availability," not less. This was a near-religious belief at the firm.[49]

So it was that Coca-Cola's vision for a sustainable future was constrained by a model for growth passed down by company leaders, decade after decade, that was all about making cold Coca-Cola available for instant consumption, anywhere and anytime. It was a strategy first developed in the rural roads of the Jim Crow American South, but one later adapted to global markets. Questioning the "arm's reach from desire" mantra, challenging that age-old mission to get cold Coke into the smallest country market, even if it meant radically reducing Coca-Cola's footprint, was clearly something that was off-limits in the environmental meetings Seabright sat in on over the years at Coca-Cola headquarters.

But fundamentally altering that model for growth could unlock remarkable opportunities for reducing greenhouse gas emissions. Do Coke cans really need to be cooled twenty-four hours a day, seven days a week, at the corner gas station in every small town across the globe? What would our world look like if the convenience of immediate consumption was sacrificed for emissions reductions critical to cooling a warming planet?

It makes sense that this will be hard work. The logic of promoting ubiquity is ingrained in Coca-Cola's corporate DNA. For more than 130 years, Coca-Cola executives have been driven by a desire to bring cold Coca-Cola to rural worlds far from city centers. Looking out from the railroad yards of Gilded Age Atlanta, Asa Candler must have felt a sense of awe at the prospects for growth in untapped markets that knew nothing of his brand. But in the era of climate change, few could argue that Coca-Cola suffers from an availability problem. Coca-Cola may not always be an "arm's reach from desire," but it's awfully close.

Coke's refrigeration dilemma, in other words, is not merely a technical problem; it's a historical one. It is the result of a commercial philosophy centered on promoting ubiquity that was born decades ago in the Gilded Age and that is no longer suitable for the world Coca-Cola inhabits today. To change the firm, Coke's sustainability officers must have hard conversations in the C-suites of the company's North Avenue headquarters. And this time, the conversation should be less about what to put into company refrigerators, but whether the company needs those refrigerators at all.

Coke is not the only southern company in need of a reckoning with its past. Another planet-changer from the American South, Delta Air Lines, is facing similar questions about the future sustainability of a business model developed in the era of Jim Crow. That story begins in Monroe, Louisiana, a small southern town where Joseph Biedenharn, the progenitor of Coca-Cola's bottling empire, located his firm's corporate headquarters just a few short years before a crop-dusting firm came to town to launch what would become the largest airline in the world.[50]

PART TWO
On Top of the World

Chapter Three

THE TRUNK LINE TO SUNSHINE

The smell of jet fuel in the morning reminds me of high school. Every day at Woodward Academy in College Park, Georgia, I heard the roar of passenger jets finishing their final approach as they landed just two miles from the center of our school grounds. Sometimes you felt vibrations while sitting in science or history class, as airplanes unfurled their landing gear, zooming toward the runway at Hartsfield International Airport, now the world's busiest airport. In 2003, the city renamed this buzzing transportation hub Hartsfield-Jackson Atlanta International Airport, to honor the city's first Black mayor, Maynard Jackson. In 2019, prior to the global disruptions of COVID-19, over 100 million people traveled through Hartsfield-Jackson terminals, which served as portals to over forty-five countries around the globe and more than 150 cities in the United States.[1]

This was the American South I knew in the 1990s. Every weekday, I commuted to school, driving down the gnarled highway junction of Interstate 75 and Interstate 85 that weaved through the center of town. I passed the World of Coca-Cola—then adjacent to the subterranean mall known as Underground Atlanta—CNN's headquarters on Marietta Street, and the cauldron that held the 1996 Olympic flame. These landmarks served as constant reminders of my hometown's link to a world beyond.

It was all connected. Hartsfield-Jackson Atlanta International Airport began as Candler Field in 1925, named in honor of longtime Atlanta benefactor and Coca-Cola founder Asa Griggs Candler. In 1909, Candler leased this land to the Atlanta Automobile Association, an organization his son, Asa Jr., had once led. The automobile group hoped to develop a racetrack there, but that endeavor soon proved unprofitable, and the association was never able to honor its debts to Candler, leading the Coke baron to turn to airplanes to bring in revenue. Nature had a lot to do with the selection of this locale for automobile and aviation exercises. Candler's property was situated in the relative flatlands south of Atlanta, a landscape in sharp contrast to the hillier foothills just north that defined Atlanta's city center. The open, level land just twelve miles from Peachtree Street, once a ridgeline Native American trail, was a prime place for planes to land, something that did not go unnoticed by mayor Walter Sims and city alderman William B. Hartsfield, who negotiated a contract with Asa Candler to develop an airport at the site in 1925. It was a generous deal, with Candler offering his property for five years to the city gratis, except for small payments to cover taxes. Thus, Atlanta's high-flying jet culture was built atop pavement and philanthropy made possible from the proceeds of soft drink sales.[2]

Like Coca-Cola, Delta Air Lines was instrumental in making Atlanta's cosmopolitan environment possible. In 1941, Delta moved its headquarters from Monroe, Louisiana, to Atlanta's Candler Field, beginning a long-term relationship that helped turn this onetime railroad hub of the former Confederacy into a global transportation gateway connecting the American South to the Global South and the rest of the world.

When Delta first came to Atlanta, it called itself "The Airline of the South," but by 1997, when I was entering my sophomore year at Woodward, the company claimed global conquest with a new slogan: "On Top of the World." The branding hinted at the company's primacy in the airline industry, and it spoke to a fundamental reality: company planes now dotted the world's atmosphere.[3]

In a very real sense, Delta had become a world changer, not only in terms of its influence on the international airline industry but also in terms of its impact on global ecosystems. By 2012, Delta Air Lines was "the world's largest commercial buyer of jet fuel," burning approximately 3.8 billion gallons annually. According to the company, compounds released during combustion of jet fuel represented roughly 98.9 percent of the firm's greenhouse gas footprint, which totaled over 38 million metric tons in 2013 and 41 million metric tons in 2018.[4]

What effect was Delta's fossil fuel consumption having on climate change? In 2022, the United Nations Intergovernmental Panel on Climate Change (IPCC) reported that overall, global transportation accounted for roughly 23 percent of all carbon dioxide emissions annually, with air traffic representing 12 percent of that transportation total. The IPCC also concluded that aviation accounted for approximately 2.4 percent of all human-generated carbon dioxide emissions. Focusing specifically on domestic and international air traffic conducted by U.S. carriers in 2005, the Pew Research Center calculated that American airlines emitted 249.3 metric tons of greenhouse gas emissions. That year, Delta reported greenhouse gas releases totaling more than 46 million metric tons, over 18 percent of the industry total.[5]

And Delta's contributions to climate change were likely to grow as this book went to press. At the turn of the twenty-first century, a substantial portion of greenhouse gases generated by the transportation sector came from moving freight. By the 2000s, less than 1 percent of all freight by weight imported or exported from the United States traveled on airplanes. Most freight tonnage in the United States moved by way of trucks on highways (58 percent), and railroads transferred roughly 12 percent of the country's freight weight. Studies in the 2000s projected that airline industries would experience considerable growth in freight traffic in the coming decades and would therefore see increased emissions. Air passenger traffic was also expected to grow exponentially. By 2050, the International Civil Aviation Organization estimated, greenhouse gas emissions from the airline industry could increase by roughly 700 percent from 2005.[6]

Thus, Delta, the largest airline in the world in 2022 in terms of revenue and total assets, was a firm that could make decisions about the future that would have Earth-changing global environmental effects.[7]

But natural resource dependencies also shaped the rise of this powerful corporation. Throughout its history, Delta faced many environmental crises. Devastating crop pests, floods, and hurricanes all left their mark on the corporation. Today, as the company faces a new environmental challenge—global climate change—history offers lessons that will prove instrumental in shaping the future sustainability of the firm. That history begins in the hot and humid river delta that first gave the airline life.

From the very beginning, nature was instrumental in designing Delta's destiny. The company owes its origins to a natural disaster brought on by monocrop cotton production in the Mississippi Delta's Black Belt. The problem was the boll weevil, a pest that both feeds on cotton fibers and reproduces inside cotton bolls. This beetle first began to wreak havoc on Texas cotton fields in the 1890s but soon migrated to monocrop fields in the Mississippi Delta, crossing the natural barrier of the Mississippi River in 1907. The outbreak drew the attention of the U.S. Congress, which established an agricultural extension service in Tallulah, Louisiana, to deal with the infestation.[8]

The insect sparked palpable fear throughout the cotton-producing South of the 1910s and 1920s. Some early reports estimated that during the boll weevil's reign, cotton growers lost "tens of billions of pounds of cotton" worth roughly $1 trillion. But beyond the material effects the boll weevil had on crops, the specter of its Sherman-like conquest of the South proved transformative as well. As environmental historian James C. Giesen put it, "It was the *idea* of the boll weevil, more than the physical destruction it wrought, that most profoundly changed the region." By the end of the 1910s, with USDA maps showing the insect threat rolling into Georgia, this beetle became a sinister pest, biblical in its proportions, that many believed threatened the very vitality of the region's economy.[9]

The boll weevil crisis became the target of Progressive Era agricultural reformers who saw an opportunity to mobilize new scientific tools to slay the American South's menace. One set of experiments, pioneered on Delta Pine and Land Company's (DPLC) Mississippi farms in the 1910s, were plant-breeding programs designed to isolate a cottonseed variety that matured faster than standard seeds, thus preventing beetles from damaging vulnerable buds during peak feeding and reproductive periods in the summer. These experiments, which led to the mass marketing of the wildly popular Express cottonseed, made DPLC into a corporate giant that one day became a critical component of the Monsanto Company's agricultural biotechnology portfolio. In other words, the birth of the life sciences seed industry had deep roots in the boll weevil days of the Progressive Era South.[10]

A seed empire was just one by-product of the boll weevil's ravenous rampage. Chemical and aviation enterprises also gained a huge boost during these years because of the South's agricultural problems. Despite the efforts of DPLC and other plant-breeding specialists, the boll weevil rolled on, forcing reformers to consider other strategies to curb beetle infestation. By that time, synthetic insecticides seemed a viable option. For decades, farmers had used chemicals to kill off unwanted pests on their land, but as late as the mid-1910s, popular mixes, such as lead arsenate and Paris green (copper arsenic), proved difficult and costly to apply in appropriate doses over large swaths of land to make a critical impact. In 1917, however, a scientist named Bert R. Coad working at the USDA's agricultural experiment station in Tallulah, Louisiana, made an important discovery. Through diligent experimentation, Coad found that calcium arsenate was a powerful pesticide that proved effective at annihilating boll weevil populations. In the months ahead, he helped create dispersal machines that could be mounted to mule-drawn vehicles to accelerate the pace with which chemicals could be spread across expansive cotton plantations.[11]

Still, there were problems with this mechanized method of dispersal—pesticides often fell to the ground or blew in the breeze—so Coad and others began to experiment with a new strategy that involved airplanes. By the 1920s, advances in aviation, spurred on by wartime investment in military aircraft, helped give birth to a nascent commercial aviation industry. One area drawing investment at the time was aerial crop-dusting, and one firm making a big play to be a leader in this field was Huff-Daland Company, a small aircraft manufacturing firm based in Orangeburg, New York. Started by Thomas H. Huff and Elliot Daland in 1920, Huff-Daland had learned of Coad's spraying campaigns in 1923. A year later, Huff-Daland secured a contract with Coad to begin aerial application of calcium arsenate in the Mississippi Delta. Signaling its commitment to this new business venture, the company created a separate division called Huff Daland Dusters that year.[12]

The government contract was a big deal for Huff Daland Dusters because at this early date airplane companies were not making much money on passenger flights. Plane engines were not powerful enough to enable aviation firms to profitably transport large numbers of passengers over long distances. As a result, companies like Huff Daland Dusters had to find other ways to generate revenue, and big agricultural contracts were one such source of funding. Servicing the countryside was the best way for Huff Daland to make money.

The promise of profits drew Huff Daland Dusters southward. Originally, the company's crop-dusting arm, led by Huff Daland vice president George Post and former army pilot Harold Harris, set up headquarters in Macon, Georgia, in part because state government officials there showed strong interest in agricultural spraying experiments. But in 1925, Huff Daland Dusters moved its headquarters to Monroe, Louisiana, just a few dozen miles away from Coad's Tallulah base. The move was encouraged by C. E. Woolman, who joined Huff Daland Dusters that year. Woolman was an experienced agricultural engineer, a graduate of the University of Illinois who had come south to work on a plantation in Louisiana. In 1913, he took a job at Louisiana State University and worked in the institution's agricultural extension service, which received a boost of federal funding after the passage of the Smith-Lever Act in 1914. Woolman spent several years as a county agent on farms in and around Monroe, and his expertise gave him broad authority to shape decisions about what path Huff Daland should take with its dusting operations in the years ahead. Considering the prospect for growth in the vast cotton stretches of farmland in and around the Mississippi Delta, it made sense that Woolman would encourage a move westward to a place he knew well.[13]

The rural South was undergoing dramatic changes when Huff Daland made its move. The strictures of Jim Crow society were tightening, as southern state legislatures strengthened laws designed to denigrate, disenfranchise, and dehumanize Black citizens in the American South. Ghastly and horrific lynchings of Black southerners continued as a resurgent Ku Klux Klan fed on a growing nativism sweeping the country to build momentum for racial violence against nonwhites throughout the Southland and beyond. Given these realities and the fact that sharecropping offered limited opportunities for economic advancements, many Black southerners living in rural communities began leaving the American South for industrial jobs in northern cities, part of the Great Migration that lasted from 1916 to the Great Depression and even beyond. Between 1910 and 1920 alone, over 10 percent of the Black population of Alabama and Mississippi left home, many enticed by headhunters for northern companies trying to fill positions opened during World War I. By the 1930s, the number of Black migrants who had left the South since 1916 totaled over 1.5 million.[14]

In this environment, chemicals, along with tractors and other mechanized equipment, appealed to elite southern white planters. These farmers still could draw on a vast population of workers by the mid-1920s, but as southern historian Jack Temple Kirby has pointed out, Black out-migration was one of the factors that pushed growers toward new labor-saving techniques, including deeper reliance on chemical pesticides. By the 1920s, growers with large, irrigated farms in places like California were already turning to chemical sprayers, tractors, and new combines to harvest their crops. Capital-starved southern farmers, on the other hand, were far slower to adopt what might be called this style of "factory farming" than their western counterparts were. It would take a massive infusion of capital delivered via New Deal programs like the Agricultural Adjustment Act of 1933 (which allowed plantation owners to use federal subsidies to mechanize their farming operations and evict tenant farmers from their land) to fully transform the face of farming in the rural South. But in the 1920s, that transition to a new tractor-driven and chemical-laden agricultural economy was already taking root in southern soil. As Black citizens poured out of an increasingly oppressive Jim Crow South, chemicals came flooding in, part of a larger shift in American agriculture toward capital-intensive farming that over the next three decades would replace the labor-intensive practices that had for so long defined southern agriculture.[15]

Woolman, Post, and Harris wanted to take advantage of this turn toward factory-style farming not just in the rural American South but also in other parts of the nation—and even the world. By 1927, the firm flew dusting planes into North Carolina and as far west as California (for pesticide campaigns in the booming and recently irrigated Imperial Valley). Southward, the company moved beyond U.S. borders, venturing in 1927 to Peru, where an opposite growing season offered Huff Daland the opportunity to secure lucrative dusting contracts during winter months in the United States. By 1928, the company boasted over twenty planes in its fleet, which historians W. David Lewis and Wesley Phillips Newton claimed was the "largest private aerial fleet in the world" at the time.[16]

Delta's expansion to Peru was in part shaped by one of the worst natural disasters in United States history. In 1927, much of the Mississippi

River lowlands in the delta and beyond were inundated with water when heavy rains led to massive levee breaches. With cotton fields destroyed, Huff Daland Dusters took a heavy hit. In 1926, company spraying contracts covered an estimated 45,000 acres, but in the wake of the disaster a year later, that number dropped to a piddling 7,800. In this moment, company officials learned the hard lesson that its fate was intimately tied to environmental realities that were in many ways beyond company control. In the wake of this disaster, Huff Daland Dusters sought new lands beyond southern borders to expand its operations, and Peru, with its growing cotton cultivation, proved an attractive global solution to a local problem.[17]

By this time, wealthy financiers in New York had taken control of the Huff Daland Company and renamed it the Keystone Corporation. These investors had become uninterested in the dusting operations undertaken by the firm's subsidiary in Monroe and were looking to unload these assets by 1928. The future of commercial aviation, these financiers believed, was not in agricultural operations but, rather, in airmail and passenger service. In 1925, Congress had passed a groundbreaking piece of legislation that legalized private bidding for federal airmail contracts. By the end of the 1920s and into the 1930s, budding airline companies, such as Eastern Air Transport (later Eastern Airlines) and Aviation Corporation (also called AVCO, later American Airlines), competed for bulk shipments of mail at rates that offered healthy profits. At the same time, advances in aviation technology resulted in the manufacture of larger aircraft with bigger engines capable of profitably carrying small numbers of passengers on short flights. Many airlines began to scale their operations up in order to offer businessmen regional service to popular commercial hubs.[18]

Whereas Wall Street turned its back on Huff Daland Dusters, Woolman saw value, and in the fall of 1928 he purchased the subsidiary from Keystone, renaming the company Delta Air Service to honor the Mississippi Delta, which had played host to the firm in its early years. The company name was suggested by Catherine Fitzgerald, a longtime administrative clerk at Huff Daland who would go on to become the first female board member of a major airline.[19]

The seed money used in the Huff Daland Duster purchase in 1928 ultimately came from seeds planted in the ground. After all, much of the financial capital Woolman raised was from local investors, such as wealthy planter Douglas Y. Smith, who purchased ten shares of the Delta

Air Service stock in November 1928 and became president of the company that year. He was joined by Travis Oliver, a banking magnate in Monroe whose financial solvency ultimately derived from the cash crops his clients grew nearby. A year later, John S. Fox, the powerful scion of a paper-mill family from Bastrop, Louisiana, came onboard as a major investor in Woolman's business. In short, extractive industries that exacted a heavy toll on the American South's forests and fields provided the financial capital that undergirded Delta's early commercial ascent. Though the firm did its business thousands of feet in the air, it was ultimately connected to the Southland.[20]

The early 1930s proved tough times for the fledgling airline. Delta historians David Lewis and Wesley Philips Newton aptly dubbed this period "doldrums" in their company history. The decade began with Delta unable to secure a lucrative government airmail contract covering territory from Alabama to Texas, a route the government ultimately gave to AVCO. Forced to reincorporate under the name Delta Air Corporation, the firm hobbled on for several years earning slim margins from dusting operations, leasing out its air hangars, and renting equipment to private plane owners.[21]

But things improved in 1934, when the company finally beat out its old rival AVCO, securing a southern airmail contract that brought in much-needed revenue. This enabled Delta to open profitable passenger service over its southern routes by the end of the decade. By 1941, the company operated relatively small planes compared to the jumbo jets of today, with a fleet that included five Douglas DC-3 twin-engine planes capable of transporting twenty-one passengers and four ten-seater Lockheed Electras. These planes serviced routes that stretched from Fort Worth, Texas, to Charleston, South Carolina, on an east–west axis as well as passenger travel north from Atlanta, the company's new hub in 1941, to Cincinnati, Ohio. As America entered World War II, revenue generated from passenger travel finally surpassed revenue coming in from airmail contracts, a remarkable transformation considering that, in 1935, the company earned three times as much money from airmail deliveries as it did from passenger service. Dusting remained a profitable enterprise for the firm, bringing in a net profit of over $9,000, but it was clear by this time that passenger transport was the key to Delta's future financial success. A company that in 1935 serviced just 4,104 passengers was now flying over 58,000 paying customers throughout the South and Midwest.[22]

Nevertheless, when the United States entered World War II, Delta was still a tiny outfit compared to other major airlines. Its early failure to secure a major airmail contract had really hindered its ability to compete with AVCO and other majors. But between 1941 and 1945, the company began to experience tremendous growth, and during that period, company assets grew from around $1.2 million to over $5.5 million. Part of this revenue boom could be attributed to military contracts that had Delta transporting personnel to bases throughout the American South, but ecological factors also boosted the airline's business.[23]

By the 1940s, many war-weary Americans sought rejuvenation in sunny climes, and Delta sought to capitalize on its access to sun-drenched regions. When the company secured federal approval for a Chicago-to-Miami route in 1945 (the "longest domestic route ever awarded" by the Civil Aeronautics Board), the company began calling itself the "Trunk Line to Sunshine." By that time, many Americans firmly believed that exposure to sunshine was essential to promoting health, especially in the wake of Progressive- and New Deal–era studies that showed exposure to sun could boost vitamin D levels. Celebrities such as Coco Chanel also valorized sunbathing in the 1920s, promoting sunbaked skin as the new, hip fashion, in contrast to the passion for pale skin that dominated the Victorian age. Companies in an assortment of industries tapped into this desire for solar radiation. For example, Sunkist and Sun-Maid advertisers developed marketing campaigns beginning in the 1920s featuring sun rays and tan women to sell oranges and raisins to an increasingly urban consumer base seeking connection to the curative powers of "rural worlds lost." The story was much the same at Delta. The company developed marketing campaigns that promised city-dwelling customers a chance to break out of the dim confines of dark skyscraper canyons in mid-twentieth-century urban America. And Delta was in a prime position to service this craze for solar rays because most of its transportation arteries stretched through the Sunbelt.[24]

But what pulled thousands of people into Delta's southern market after World War II was as much the economic climate as the ecological climate. Postwar military investment in places like Jacksonville, Florida, Oak Ridge, Tennessee, and Huntsville, Alabama, meant there were now new centers of business seeking connection to a larger world. Considering that air travel was still an expensive affair, Delta depended on markets where corporate investment was flourishing. The company simply could not have survived on tourism alone. Thus, the migration of industry and personnel from the Midwest to the South and Southwest was critical to keeping the company's planes filled in the 1940s and 1950s.[25]

This migration, however, might not have happened without dramatic man-made alterations to Sunbelt ecosystems in the postwar era. Sun was not always the American South's best-selling point, especially in the particularly humid months of the summer. Southern historian Raymond Arsenault explained how the rapid and widespread expansion of air-conditioning infrastructure in Cold War America enabled businesses to overcome a natural barrier to development in the American South. Air-conditioning created a business and residential climate that was more appealing to people considering moving to such a hot region of the country. This was another boon for the southern economy that redounded in Delta's favor.[26]

In places like Florida and the lowlands of Louisiana, the postwar use of DDT, a pesticide developed in wartime to protect soldiers from typhus-carrying fleas and malaria-laden mosquitoes, helped the South shuck off its long-held image as a diseased landscape. In this case, the forces shaping Delta and Sunbelt ecosystems worked bidirectionally, because it was the airmen working for Delta in the 1940s that had pioneered the aerial dusting techniques that would be used to deploy DDT throughout the South during malaria eradication campaigns. As Delta executives boasted in 1943, the airline had "dusted poison by airplane for the control of mosquitoes and the prevention of malaria in territories adjoining or in the vicinity of military camps in various southern states." Thus, Delta had a direct role in reshaping the insect-laden environment of the American South.[27]

These campaigns did not eliminate malaria in the Sunbelt. Some scholars have argued that the greatest strides in regional malaria abatement should be attributed to New Deal health programs initiated in the 1930s. Others hold that southerners' migration from rural farmland played a pivotal role in reducing incidences of malaria in the region. But in any case, many who came south in the postwar period surely believed DDT-spraying campaigns contributed to reductions in malaria infection rates in the American South, which in turn made Delta's prime areas of operation appear more attractive for residential and commercial development.[28]

Finally, Delta's growth in this period also hinged on federal aid that improved the built environment of southern airports. In the 1940s and 1950s, air facilities in Atlanta, Memphis, New Orleans, and other southern towns received significant federal funds to build out their infrastructure. Under the Federal Airport Act of 1946, airport authorities could secure federal grants from the Federal-Aid Airport Program (FAAP) to help construct terminal buildings, runways, radar towers, and other infrastructure. This capital proved critical for southern airports that needed major overhauls to take advantage of the growing traffic at their terminals.[29]

But some in Congress threatened this critical supply of funding in the mid-1950s, asking whether the Civil Aeronautics Administration (CAA), which administered the federal aid program, should offer funds to hubs such as Atlanta's airport given that these transportation centers were racially segregated. Though the Civil Aeronautics Act of 1938 prohibited airline carriers from segregating seating on planes or discriminating based on race, this law did not specify that airports had to be integrated facilities. As such, officials receiving FAAP monies in southern cities often used funds to finance segregated waiting areas or duplicate restrooms for Black and white customers. Not only was this wasteful, some legislators charged, but it also seemed at odds with the recent *Brown v. Board of Education* (1954) decision overturning the *Plessy v. Ferguson* (1896) doctrine, which had allowed public accommodations to be segregated so long as facilities were "separate but equal."[30]

The CAA ultimately decided never to directly challenge segregation in airports; instead it settled on a policy that merely prohibited the specific earmarking of funds for use in expanding or constructing spaces intended for segregated use. By 1960, the CAA adopted new rules that said FAAP funds could only be used for operations critical to airport safety, which meant southern airports were allowed to maintain segregated waiting rooms, restrooms, and restaurants on their own municipal dime without fear of jeopardizing funds that could be used elsewhere in the airport. In short, the CAA looked the other way when confronted with Jim Crow segregation, benefiting southern airports, which were still able to receive federal funds to build out their facilities.[31]

Delta executives, for their part, quietly complied with segregated airport facilities, though they vowed not to discriminate against their paying customers while in their care. When Congressman Charles C. Diggs (D-Mich.) launched an investigation into segregated airport facilities in 1955, Delta's C. E. Woolman responded by saying that Delta did "not encourage the practices of which you complain" and added, "Where local laws require separate facilities . . . we do make every effort to insure the provision of equal and acceptable facilities by the governmental owners of the airport properties."[32] Woolman was clearly saying his firm would honor local ordinances that segregated passengers in airports so long as facilities were "separate but equal," even in a post-*Brown* world.

By the early 1950s, Delta Air Lines (so named in December 1945 because of the company's remarkable passenger service growth) had become one of the biggest airlines in the country. During the decade, the company bought out competitors, such as Chicago and Southern Air Lines, and vastly increased its market reach, which now included international airports in the Caribbean. By 1956, the company had finally secured access to New York and the bustling metropolises of the Northeast and Mid-Atlantic. The company also worked with American Airlines on an interchange service to California. With bigger planes, including forty-four-passenger Convair 340s and sixty-nine-passenger DC-7s, Delta's contrails now crisscrossed the country.[33]

But what remained so notable about this high-flying firm in the 1950s was the way it continued to service smaller towns in more rural parts of the segregated American South—the places that had first given the firm life. Some Delta routes certainly terminated in cities like Chicago (population 3.6 million in 1950) and Los Angeles (1.97 million) by the mid-1950s, but the firm still picked up passengers in Monroe, Louisiana (population 38,572 in 1950), and Meridian, Mississippi (41,803). Delta's eastward flight artery, the one that had helped get the company off the ground, swung straight through the Black Belt of the American South, with stops in towns like Selma, Alabama; Columbus, Georgia; and Spartanburg, South Carolina.[34] Delta, in other words, may have been doing big international business by the 1950s, but it was still very much a southern firm, and not just because it drew customers from small southern towns. It also depended on critical resources extracted from the Southland to keep its planes in the air.

Chapter Four

FUEL PROBLEMS

In its earliest days, Delta benefited from a petroleum boom that energized America's increasingly motorized economy. In the 1930s, Delta's planes burned high-octane gasoline (73-octane and 80-octane—higher than the 63- and 64-octane gasoline used by cars during the Great Depression). At that time, gasoline production was growing exponentially, in large part because of booming oil production in Texas, Oklahoma, and California and the rapid expansion of refinery infrastructure in the United States.[1]

Delta's birth coincided with healthy competition in the oil industry. Rockefeller's Standard Oil Trust had been broken up in the 1910s, meaning the southern airline could negotiate contracts with different oil outfits each competing for the company's contracts. Records preserved in the firm's corporate archive reveal that Delta negotiated with Shell, Texaco,

Standard Oil of New Jersey, Gulf Refining Company, Sinclair Refining Company, and others throughout the 1930s.[2]

Nevertheless, fuel was one of the biggest expenses for Delta's passenger service. A handwritten calculation of projected costs for Delta's two-way service between Birmingham, Alabama, and Dallas, Texas, in the late 1920s showed monthly gas costs around $1,330. Besides "depreciation and maintenance," only the costs of "four pilots and service," at $1,375, topped fuel expenses. As such, C. E. Woolman worked hard to get the best prices for gasoline, and petroleum companies, for their part, had incentive to make attractive offers. Contracts could top 500,000 gallons annually, which could mean sizable revenue for oil refining firms.[3]

There was also the issue of petroleum transport, which limited how much fuel the company could channel into its corporate system at any given time. Today major airports, such as New York's JFK and Atlanta's Hartsfield-Jackson International Airport, have pipelines that connect them to petroleum refining plants hundreds of miles away, but back in Delta's fledgling years, gasoline had to be transported to most airports via railways and trucks. As Standard Oil of New Jersey's chief aviation officer explained to C. E. Woolman in 1935, oil companies could utilize tank cars running on rails to service Delta's hubs in Dallas, Texas; Jackson, Mississippi; and Atlanta, Georgia, but smaller airports in rural Monroe, Louisiana, and Augusta, Georgia, necessitated the use of "tank wagons." Delta's business could only move as fast as the railcars and trucks that brought it fuel.[4]

Nevertheless, American infrastructure for aviation fuel production and transport expanded dramatically in the next three decades, with World War II being a particularly transformative moment. In January 1942, just a few weeks after Pearl Harbor, the *Wall Street Journal* reported that the federal government had "nationalized" the "production and sale of aviation gasoline" that was higher octane than earlier fuels and called "avgas." The government put rationing schemes and price ceilings in place to ensure that America's flying servicemen had the fuel they needed to defeat Japanese and Nazi adversaries. The War Production Board and the Office of Price Administration also initiated motor gas conservation programs, something C. E. Woolman noted in a 1943 letter to Delta workers. He congratulated "all Delta employees who are now participating in share-a-ride groups, who are using bicycles and motorcycles in fair weather; and I wish to encourage every possible economy in the use of gasoline by Delta employees."[5]

Yet even as Woolman spoke of conservation to his employees, Delta witnessed exponential growth in its consumption of aviation fuel during World War II. Partnering with the U.S. military's Air Transport Command, Delta shipped American GIs and their cargo across the country, clocking in 2.5 million miles of air travel and over 6.6 million passenger miles for Uncle Sam. On the civilian front, wartime investment in the American South boosted many southerners' incomes dramatically, which fueled even more Delta traffic as businessmen traveled from burgeoning commercial hubs throughout the wartime South. The result was a dramatic increase in air traffic. Between 1941 and the end of 1944, the company upped its total passenger miles (number of passengers multiplied by number of miles) from just over 36 million to roughly 57 million.[6]

Massive investments in oil refining capacity fueled all this growth. Even before the United States' official entrance into war, total aviation fuel production increased from roughly 25,000 to 50,000 barrels a day in 1941 alone. Based on careful analysis of business trends, the *Wall Street Journal* predicted in 1942 that capacity would grow to over 120,000 barrels by 1943, nearly a fivefold increase in just three years. In the postwar era, Delta fed on this improved petroleum production infrastructure to expand its commercial empire.[7]

Fuel and oil remained the most expensive single-item operating costs on Delta's books besides wages and salaries by the 1950s, fluctuating between 10 and 15 percent of total operating costs. Nevertheless, executives engaged in little discussion about fuel efficiency in corporate annual reports during this period.[8]

In 1959, Delta, still under the leadership of its visionary founder C. E. Woolman, entered the jet age with the purchase of its first DC-8s, and by the early 1960s the company had added eighty-four-seater Convair 880s to its fleet. The future of commercial flight was in jet-propelled planes, not the old piston-powered aircraft of the past. These Convair planes could cross the country at top speeds over 600 miles per hour (more than twice as fast as the DC-3s popular in the early 1940s). The DC-8s could carry 119 passengers at similar speeds. In economic language, Delta had found a way to dramatically increase its throughput—that is, the flow of passengers through its corporate system—which resulted in increased profits.[9]

One key to Delta's success at this time was its "hub-and-spoke" flight system. Adopted in 1955, long before the other major airlines embraced the idea, Delta's aviation plan involved flying planes from a variety of smaller southern towns to the airlines' central hub in Atlanta, where passengers could then be fed to aircraft leaving for longer flights. "We were the first airline to develop the hub-and-spoke system," one Delta executive recalled, noting that studies had indicated that this was "the only way that [Delta's] system would make money."[10]

Delta had long serviced small southern towns, including Macon, Georgia; Monroe, Louisiana; and Hattiesburg, Mississippi, and now, with this hub-and-spoke system, it had found an efficient and cost-effective way to connect passengers from these less-urban areas to a national transportation network that stretched from coast to coast and from the American South all the way to New England and beyond. Finding ways to link small southern cities and towns to Atlanta proved the path to profits for Delta.[11]

Even though Delta took passengers farther and faster in the 1950s and 1960s, the company did not see substantial changes in their fuel costs as a percentage of total operating expenses. In fact, in 1961, the company reported fuel and oil costs around 12 percent of total operating outlays. This was less than the company had reported in the mid-1950s.[12]

How did the company pull this off? The trick was that jetliners used refined kerosene rather than the much more expensive high-octane gasoline. New jetliners used more fuel than their propeller predecessors, but because airlines could purchase kerosene-based jet fuel at lower prices than aviation gasoline, the cost difference was marginal.[13]

Cheap jet fuel continued to spark Delta's exponential growth throughout the 1960s, as did new infrastructural investments, especially the construction of new petroleum conduits such as the Colonial Oil Pipeline (completed in the 1960s by a Georgia firm). Extending from the oil-rich regions of Louisiana and East Texas, the Colonial Pipeline featured branch lines that fed Atlanta's airport and other major transportation hubs up and down the East Coast. No longer was Delta solely dependent on slow, tank-car deliveries common in the 1930s and 1940s. Now fuel came flooding into airports direct from refineries hundreds of miles away. The airlines' soaring growth was in part a product of a fuel bonanza in the American South and Southwest as commercial arteries fed refined petroleum into bustling cities.[14]

Cheap oil helped fuel Delta's growth by the 1960s, but so too did something else: the slow, but steady dismantling of de facto segregation in the Jim Crow South. By the early 1960s, a sit-in movement swept the nation, led by young college students in Greensboro, Nashville, and elsewhere, and the Freedom Rides of 1961 set off from Washington, D.C., destined for the Deep South. By 1963, the visual spectacle of Birmingham's police chief Eugene "Bull" Connor turning a fire hose on children protesting in Birmingham brought national attention to the plight of Black citizens in the American South.

Delta's Atlanta airport hub was one site of protest during this heyday of civil rights campaigns. In 1959, an interracial group of activists associated with the Congress of Racial Equality (CORE) demanded to be seated together in the segregated Dobbs House restaurant in the Atlanta airport. Less than six months later, CORE and other civil rights activists again tried to integrate airport facilities in Greenville, South Carolina, another frequent stop on Delta's southern trunkline. Then, in 1961, "Freedom Flyers" boarded a Delta plane destined for Jackson, Mississippi, where the group sought to challenge that airport's Jim Crow policies. All this activism led the Justice Department to get involved, officially ending de facto segregation in southern airports by 1963, a year before the passage of the Civil Rights Act.[15]

Delta officials did not take an active stance in fighting for this change, but the firm's planes proved critical in speeding up justice for Black southerners. Reverend Martin Luther King Jr., Ralph D. Abernathy, and other civil rights leaders were frequent Delta travelers, hopping from one major event or protest to another. And there were also other professionals—especially members of the media—who were able to swiftly fly film crews to broadcast scenes of bloody conflict to a national audience. This flood of information, enabled in part by Delta's aviation network, helped to bring down a Jim Crow system that had been holding southern states back economically for so long.[16]

Though Delta did little to proactively spur the civil rights reform that swept through the South in the 1960s, the airline nonetheless benefited dramatically from the civil rights movement. The southern economy boomed after the collapse of de facto Jim Crow segregation in the South, as businesses that had once been shy of investing in a region so mired in racial strife came south. As southern historian Edward L. Ayers put it, the "moral tariff" of segregation had been lifted. The 1950s and 1960s had seen millions of citizens, Black and white, migrate out of southern states, but that trend changed after the passage of the Civil Rights Act of 1964 and the Voting Rights Act of 1965. Nearly 3 million people migrated to southern states between 1970 and 1976, including many Black Americans who still had family and connections in the region. Here was a flood of new customers Delta was eager to serve. It took time for white southern businessmen to recognize that integration was good for business, but in time they were convinced by rising retail sales and a flood of new investments to accept the new social order that benefited their bottom line.[17]

Delta, now the dominant airline in the fastest-growing region of the country, was ready to make big profits in this new commercial ecology of the post–Jim Crow South. But global political and economic forces had other plans.

In 1974, for the first time in the company's history, Delta featured a section titled "Fuel Problems" in its annual report in which it told shareholders the company was facing serious cost increases due to fuel price hikes. The problem was a temporary peak in domestic oil production coupled with the Organization of Petroleum Exporting Countries (OPEC) oil embargo of 1973–74. Facing national shortages of fuel that had Americans waiting in gas station lines that stretched several city blocks, the Federal Energy Administration began rationing fuel. Between June 1973 and June 1974, Delta witnessed an 86 percent increase in its fuel prices, and it seemed trouble was bound to continue into the foreseeable future.[18]

Delta responded to these natural resource shortages by adjusting its business practices. It adopted fuel conservation initiatives such as "reducing cruising speeds," improving "ground handling procedures," implementing computerized flight simulator systems for new pilots, adopting flight patterns that maximized fuel use, and selling off some gas-guzzling planes and replacing them with more fuel-efficient aircraft, such as 727-200s. Collectively, these initiatives resulted in a "9% increase in fuel efficiency" between 1973 and 1974, according to the company.[19]

These adjustments, however, did not solve the problem. By 1979, when the Iranian revolution resulted in reduced U.S. oil imports, Delta once again became alarmed about fuel prices. By this time, Delta had become a transatlantic airline, having secured a direct route from Atlanta to London from the Civil Aeronautics Board in 1977. The company was expanding globally, and it needed more resources than ever to service its international air network. But foreign affairs were getting in the way. In 1979, Delta president David C. Garrett Jr. and chairman W. T. Beebe explained that the energy crisis was hitting the company hard, noting that every one-cent increase in fuel was costing the company over $12 million in additional annual expenses.[20]

Once again, the company recommitted itself to resource conservation. Delta invested in better airplanes such as Boeing 727-200s, developed flight simulator programs, and adjusted flight speeds. Yet even if the company was becoming more efficient, it was nevertheless using more and more jet fuel each year. Company officials noted that Delta used around 900 million gallons of fuel in 1974, but by 1979 total consumption was roughly 1.1 billion.[21]

Politics also gave Delta executives headaches in 1978, when Congress passed the Airline Deregulation Act, which opened the door for budget airlines to compete with the major carriers. Up to the 1970s, the Civil Aeronautics Board (CAB) had carefully managed the airline industry and regulated fares and controlled new entries into the market. This act did away with such federal management, creating a more competitive system that threatened to drive ticket prices downward. What would this do to Delta's bottom line?[22]

In this period of uncertainty, the jet fuel crisis worsened. In 1980, the company reported prices hovering around 75 cents per gallon, up from a little over 10 cents per gallon in 1973. At the start of the new decade, fuel expenses represented a whopping 30 percent of operating costs, a far cry from the 10 percent enjoyed in the good years of the 1950s. The company doubled down on conservation practices initiated in the 1970s. In 1981, the company announced the purchase of sixty "highly efficient" Boeing B-757-200 aircraft that included new technology to optimize fuel consumption. With fuel prices topping one dollar per gallon, the company felt it had no choice but to invest in aircraft that used less jet fuel.[23]

But as the recession of the early 1980s ended, the dark days of the energy crisis were over. By 1984 Delta's financials were improving, in large part because a substantial drop in petroleum prices reduced overhead costs. In 1987, Delta purchased jet fuel for half what they had paid six years earlier. The company closed out the 1980s posting $460 million in profits.[24]

Delta also benefited from federal assistance at this time through the Essential Air Service (EAS) program. Established by the Airline Deregulation Act of 1978, the EAS essentially provided subsidies to carriers servicing airports in particularly remote or rural portions of the country.

The government did this to try to ensure that major carriers did not abandon smaller airports for larger ones after restrictions on flight paths were loosened under the 1978 law. Delta, which still served numerous small towns across the American South and beyond, ultimately benefited from hundreds of millions of dollars spent through the EAS program over the next several decades which enabled Delta to continue flying to and from airfields located in less-urban parts of the country. While statistics ultimately showed that the EAS program failed at preventing carriers from abandoning some small towns, Delta nevertheless fed on these federal funds in markets where it often found it hard to fill its planes to capacity. In 2011, for example, Delta was still relying on EAS funds to justify keeping two flights a day operating from an airport in Muscle Shoals, Alabama, even though these flights were typically only 35.7 percent filled. All told, the airline benefited from $1.7 million spent by the EAS program annually just to keep these two flights operational.[25]

Back in the 1980s, Congress did not appropriate as much money to the EAS program as it would after the mid-1990s, but EAS nevertheless was useful to a firm like Delta that remained so connected to small-town America. In that decade, Delta executives contracted with small independent airlines, such as Atlanta Southeast Airlines, Comair in the Midwest, and Rio Airways in the American West, agreeing to share the Delta name with these smaller companies so that they could serve smaller markets. In a manner similar to Coca-Cola's bottling network, Delta's franchise system enabled the Atlanta aviation firm to maintain a strong presence at smaller airports across the country.[26]

But in an era of climate change, one had to wonder whether the federal government's EAS program subsidizing airline service to many of these smaller airports was good for society. Flying jets filled to less than half of capacity short distances was not a particularly efficient way to burn fossil fuels. But for a firm like Delta, EAS was a beneficial program that subsidized the company's operations in small markets.

In the 1990s Delta saw dramatic growth in its international business that was in part driven by an influx of new migrants to the American South. After the Olympics came to Atlanta in 1996, Delta experienced a surge in international travel. Between 1997 and 1998 alone, Atlanta's airport witnessed a 22 percent jump in international flights. Delta was expanding routes to Latin America, Africa, and Asia, which made sense, given that the American South had become home to many immigrants after the passage of the 1965 Immigration and Nationality Act, which loosened limits on immigration from particular countries.[27] Though this law specifically limited immigration from Mexico and other Latin American countries for the first time, Latinx people migrated to the American South in the 1970s, 1980s, and 1990s in search of work and new opportunities in a region that was economically booming. A 1986 federal law designed to give amnesty to undocumented Latinx immigrants living in the United States had the unintended consequence of encouraging many undocumented workers to stay in the region in the years after the law's passage. Migration to the American South continued as cities like Atlanta and even smaller towns in the Southeast saw rapid rises in their Latinx population. And other people were also coming to the region. By 2000, nearly 30 percent of all Asian immigrants entering the United States set up residence in the American South, as did Nigerian expatriates who traveled to cities like Houston and Atlanta in search of new educational and professional job opportunities in these fast-growing metropolises. These newcomers to the American South helped justify Delta's expansion into global markets in the 1990s and beyond.[28]

But to continue its aerial conquest, this corporate giant desperately needed fossil fuels. Fuel price volatility had rocked the company in the last three decades, forcing it to undertake conservation initiatives that drastically increased the efficiency of its fleet. But it was fossil fuel volatility of another sort that sent shock waves through Delta's corporate system in the new millennium. On September 11, 2001, over 20,000 gallons of jet fuel burst into flames as American Airlines flight 11 crashed into the Twin Towers in downtown Manhattan, followed by United flight 175. Soon thereafter, American flight 77 hit the Pentagon. Nearly 3,000 people lost their lives.

Government-imposed groundings coupled with customer fears about air travel hit the airline industry hard. All this happened in the wake of the dot-com bust, a financial fallout sparking a recession that only made matters worse. By the end of the year, the company posted a loss of more than $1 billion. Three years later, the company was still struggling, with losses over $5 billion.[29]

And then nature delivered another blow. In August 2005, Hurricane Katrina pummeled the Gulf Coast, doing significant damage to oil refineries in the region. Crashing waves and floodwaters reduced jet fuel production in the area, constricting supply by about 10 to 15 percent. This was particularly bad news for Delta because its southern base in Atlanta was connected directly to Gulf of Mexico oil reserves via the Colonial Pipeline. As oil prices skyrocketed, Delta simply could not handle the financial blow. It filed for Chapter 11 bankruptcy in September 2005.[30]

There was a great irony in all of this. Delta's explosive growth in jet fuel consumption meant that it was creating more and more greenhouse gases, which many scientists argued were contributing to climatic changes that made hurricanes in the Gulf states more powerful and destructive. Delta had long played a role in shaping the ecology in and around the Mississippi Delta region it once called home. It had helped to suppress the devastating boll weevil and sprayed DDT throughout the American South in an attempt to create mosquito-free environments that would attract families to come live in the region. But in the 2000s, the company was now contributing to global environmental changes that struck at the core of the firm's ability to make money. It was shaping the local ecology of its regional base, which in turn would affect the firm's ability to reach global destinations far beyond its home borders.

Considering this reality, it might have seemed probable for a restructured Delta to come out of Chapter 11 bankruptcy with a renewed commitment to curbing the fossil fuel addiction that had caused it so much trouble in recent years. Instead, Delta went another direction, and became even more wedded to petroleum production.

In 2012, Delta integrated into ownership of an oil refinery just outside Philadelphia, Pennsylvania, that had recently been mothballed by ConocoPhillips. The goal was to secure insulation from a volatile fuel market. Heading into the refinery purchase, Delta had announced that jet fuel costs were the single largest expense in its operating budget. This was due in large part to skyrocketing oil prices during the 2008–9 recession, which cut deep into the airline's profits.[31]

Delta's purchase was unusual. Company president Edward H. Bastian told reporters, "No airline has ever purchased a refinery in the long history of the airline industry." This was going to be a massive operation capable of providing roughly 80 percent of the company's fuel needs in the years ahead. Beyond the refining plant, Delta also acquired pipelines capable of swiftly funneling fuel to Delta's fleet departing from the bustling airport terminals in New York City, a place notorious for fuel shortages brought on by competitive jousting for scarce resources in a high-demand market.[32]

Several financial analysts thought this was a dumb move, including Edward Hirs, a University of Houston professor of energy economics, who said the only reason he could imagine Delta executives would make such a deal was "because they're stupid." As Hirs explained in 2012, Delta was taking on a tremendous amount of risk. Why not instead try and negotiate futures contracts with independent refiners using loaned money, an established practice known as "hedging" that was quite popular in the industry? As Hirs explained, hedging would limit the company's exposure to environmental disasters like oil spills or labor troubles.[33]

In the end, Delta ignored such advice and made the deal. This was not a company that was radically rethinking its dangerous dependence on fossil fuels. The driving logic at the firm was to find more oil. It was a mantra not too dissimilar to the one espoused by Gilded Age oil barons over a hundred years earlier. "The more we can maximize jet fuel production," remarked Bastian, "the more" Delta's refinery acquisition "makes sense."[34]

With its new oil refinery, Delta could literally shape commodity markets. In 2016, for example, Delta issued a company memo that explained its strategy of dumping cheap jet fuel into the New York market to reduce Delta's operating costs. Jeff Warmann, the author of the memo and the manager of Delta's Philadelphia refinery, admitted that this "negatively impacts our refinery economics" but argued that it would ultimately "greatly . . . reduce Delta's fuel costs." Delta was helping to keep cheap fuel flowing to the airline industry, which in turn reduced incentives for major carriers to pursue greener fuel sources.[35]

As petroleum prices began to come down in the 2010s, driven in large part by oil exploration in the Bakken reserve and other newly discovered shale deposits, Delta's fossil-fueled future began to resemble its past. The company started investing in older planes that first emerged in the 1970s. Economics explained the move. In 2016, Delta could purchase newer Boeing 777s for around $300 million per plane, compared to the $14 million price tag for much older 757s that ran on technology first developed in the 1970s. The 777s are far more fuel efficient than the 757s and thus produce significantly fewer greenhouse gases in flight. Aviation journalist and *Condé Nast Traveler* contributor Clive Irving reported in 2016 that Delta had "one of the oldest fleets among American carriers": many of its jets first took flight at the end of the 1990s, another period of cheap oil prices. "Forget futuristic green jets," Irving's headline read. "You could be flying old gas guzzlers for years to come."[36]

By 2016, Delta had the largest fleet of 757s of any airline, with over a hundred such models in the sky. As a result, it was also one of the biggest jet fuel consumers in the business, with fuel efficiency increasing by less than 4 percent between 2008 and 2016. Compared to the nearly 30 percent gain Delta made in fuel efficiency between 1980 and 1988, this change was minuscule.[37]

Here was yet another great irony: Delta had first chosen to invest in 757s in the 1970s mainly because these planes featured some of the best fuel-efficient technology of the day. At the time, the company was facing rising jet fuel prices and turned toward conservation initiatives to deal with the problem. Now, in an era of relative fossil fuel abundance, the firm purchased 757s instead of newer planes that were far more fuel-efficient, because, at least for the moment, it made a lot of sense for the company executives when they looked at the bottom line.

But history made clear that this was a dangerous path to follow. By the 1970s, Delta's fate was tethered to the oil market. Fuel price fluctuations, in part brought on by climate change, threatened the airline's financial viability. Efforts to improve efficiency never really solved this problem, because even though Delta became more efficient at using jet fuel, it consumed more and more petroleum over time. In 2016, the company topped the 4-billion-gallon mark, a statistic that reflected not only the company's growth in sales and attendant profitability but also its long-term vulnerability. Now Delta, bigger than ever, was going to find it harder to insulate itself from the vagaries of the petroleum market.[38]

The burning of all that fuel contributed to substantial greenhouse gas releases. Even as the company professed a commitment to combat climate change, Delta acknowledged that its carbon dioxide emissions continued to rise, from just over 38 million metric tons in 2013 to roughly 40 million metric tons in 2016. To deal with this growing problem, "efficiency" remained the choice word of the day. The term was used thirty-three times in Delta's 2016 *Corporate Sustainability Report*, evidence that the firm believed efficiency was the key to its salvation. Despite having the oldest fleet in the sky, the company heralded the purchase of new Airbus A321s and Bombardier CS100s, which would help the company reduce its fuel consumption. In just a few years, the firm promised, the company would be "traveling lighter in the skies."[39]

This emphasis on efficiency was quite familiar. In the 1970s, the company hawked efficiency as its best strategy to deal with its petroleum problem, but in time it became clear that a fuel-efficient fleet made Delta heavier, not lighter. In the four decades after Delta commenced fuel conservation initiatives, the firm witnessed a fourfold increase in its fuel consumption. Delta made it appear that investment in these new fuel-efficient technologies was a forward-thinking strategy that would radically improve the firm's environmental footprint, but Delta's plan for the future reflects old ways of doing things. The firm was investing in a strategy that was now almost half a century old and that, as history shows, did not work to solve the problems it was designed to fix.

And it could afford to do so in large part because fuel was cheap. The fracking boom of the 2010s had brought a flood of cheap fuel into the market, leading to reduced costs for the firm. This contributed to a "banner year" for Delta in 2019: the firm posted record revenue generation of $47 billion that contributed to $4.7 billion in profits.[40]

Everything changed in 2020, when the global coronavirus pandemic swept the globe and airline companies saw huge financial losses. Once again nature had intervened in Delta's corporate history, and this intervention might have undone the firm but for the help of the federal government. In 2020, Congress passed the Coronavirus Aid, Relief, and Economic Security (CARES) Act and the Consolidated Appropriations Act, which funneled $25 billion and $15 billion, respectively, to passenger airlines. In 2021, Congress also passed the American Rescue Plan, which sent an additional $14 billion to air carriers like Delta to keep them from going into bankruptcy. Drawing on this critical aid, Delta weathered the pandemic, and by the end of 2021 it was once again posting profits. Rising fuel costs associated with Russia's invasion of Ukraine were ultimately offset by surging sales in the spring of 2022 as people returned to air travel after two long years of postponing trips. Delta simply increased its ticket prices, passing along the cost of rising fuel prices to customers who were eager to begin air travel again.[41]

In these years, Delta once again told consumers it was trying to address climate change by investing in more efficient aircraft. It also said it was buying sustainable aviation fuel (SAF) derived from plant-based products such as sugarcane and waste products such as animal fats. But for anyone following this SAF trend closely, it was clear there were real problems. SAF supplies were extremely limited in 2022, and according to *Bloomberg*, some of these fuels also risked "generating more carbon dioxide than the conventional fuels they replaced." (Consider, for example, the fossil fuel inputs needed to create the biomass used for SAF fuels in the first place.) And in any case, Delta only promised to replace 10 percent of its refined jet fuel with SAF by 2030. Such a tepid commitment to sustainable aviation made sense for a firm that was directly invested in oil refining.[42]

In short, Delta never made a radical commitment to dealing with climate change because it never really faced financial pressure to do so. In many ways, despite all the chaos of the coronavirus years, Delta was getting back to business as usual in 2022.[43]

Or so it seemed at first glance. In fact, Delta was beginning to initiate some pretty substantial changes in its operations, though not with an eye toward environmental sustainability. The big change was in its flight patterns. The company was pulling out of some of the more remote and small-town airports it had long served in the decades before the pandemic. The company cited the financial difficulties of the COVID crisis as one of the primary reasons the firm was doing this, but Delta had actually started talking about cutting out service to many small-town airports as early as the 2000s, arguing that even with Essential Air Service subsidies from the federal government, the company was finding it hard to justify sending planes to these smaller towns when they could make more money in metropolitan markets. The pandemic, in other words, accelerated Delta's shift away from smaller towns that had once been central to its hub-and-spoke growth.[44]

But Delta didn't abandon its roots completely. When the company announced the airports it would no longer serve in June 2020, only two out of the eleven locations were in the American South. Delta would no longer fly to Flint, Michigan (population 94,968), or Peoria, Illinois (109,428), or Erie, Pennsylvania (95,077), but it would continue travel to small southern towns that remained at the center of a global economy the region had helped create. One of those towns was Bentonville, Arkansas, population 57,537, home to another southern business that half a century earlier had initiated a retail revolution that changed life on planet Earth.[45]

PART THREE

Walmart World

Chapter Five

OZARK-CAJUN "BONE-MENDING" RIVER STEW

The scene would have fit well in the opening of the stereotypically southern 1972 river adventure flick *Deliverance*. Sam Walton, founder of Walmart and one of the richest men in the world, sat perched in the canoe as he helped direct it through the waters of Sugar Creek just outside Bentonville, Arkansas, Walmart's hometown. Walton, wearing his signature crumpled trucker's cap, was old hat at this. Growing up in Missouri, he had perfected his canoeing skills en route to becoming, according to a close friend of his, one of the youngest Eagle Scouts in Missouri history. Walton loved the outdoors, and so did his wife, Helen, who was also in the boat, readying herself to assist.[1]

This was the setting for the 1981 Walmart shareholders meeting—a strange place for such an affair, but that was the idea. Sam Walton loved the thought of getting these Wall Street analysts from up north out into an Ozark wild that he called home. Those who planned the outing played up the exoticism to the hilt. Visitors from out of town had been handed flyers detailing an "Ozark-Cajun 'Bone-Mending' River Stew" that Sam's daughter Alice would apparently provide for their journey. The ingredients? A mix of snakes and chicken parts brewed in a pot, tended by an overalls-bedecked heiress puffing on a corncob pipe.[2]

By all accounts, it was a raucous affair. Many of the Wall Street types got sloppy drunk, even to the point of toppling out of their canoes, which bothered Sam Walton, who henceforward made shareholders meetings a dry event. ("They were never quite the same after that," said one Walmart investor.) At night, the group camped, and chaos ensued as city slickers tried to cope with the sounds of wilderness. "That was a real fiasco," Sam Walton said. "A coyote started howling, and hoot owls hooting, and half of these analysts stayed up all night around the campfire because they couldn't sleep. We decided it wasn't the best idea to try something like this with folks who weren't accustomed to camping on the rocks in sleeping bags."[3]

The press picked up on this idea that the ecosystem Walmart called home was something foreign to the captains of industry in the Northeast. Business reporters loved to play with the contradiction that such a "down-home" company from a seemingly backward "small town tucked away in the northernmost corner of Arkansas" could have a balance sheet that was "pure sophistication." They talked about the float trips and barbecues as a staple of the business's headquarters, which seemed to stand in stark contrast to the cold calculation of computer-age capitalism. As historian Bethany Moreton explained, some people just could not understand how "high-tech rednecks mastered cybernetics and corporate culture without losing Christ or country music."[4]

But for Sam Walton, the Ozarks had always seemed the perfect environment in which to do business. Despite the failings of the 1981 shareholders meeting, Walton continued to use the Arkansas woods as his boardroom for big-deal meetings. In 1987, for example, Walton journeyed down another Arkansas waterway, this time with Procter & Gamble's vice president of sales, Lou Pritchett. Pritchett had proposed the idea because Walton, like himself, had been a Boy Scout. After forging a bond of friendship with the secret Eagle Scout handshake, the two were off, floating down the South Fork of Spring River, discussing how they could revolutionize retail. Pritchett later recounted the story of this journey in a book called *Stop Paddling and Start Rocking the Boat* and spoke of the rugged bus ride that got them between two portage points on the river. "Thick clouds of red dust poured in on us," Pritchett recalled, but the two were unfazed. When they got back on the water, Walton, with his little yellow notepad, jotted down ideas from their conversation as he held on to the gunwale of Pritchett's canoe.[5]

The river proved a fitting setting for a conversation that was all about flow. What they devised on this trip was a better way to get goods from producers to consumers: they resolved to develop a computer system that would allow for open sharing of inventory and point-of-sale information between suppliers and retailers. The system, which became known as Retail Link, would allow companies like Procter & Gamble to instantaneously restock items sold in stores. Walmart and Procter & Gamble were essentially removing dams that separated suppliers from retailers further downstream, allowing a flood of goods to course through global commercial arteries.[6] The decision changed Walmart—and then the world—forever.

But the ecological roots of the Walmart empire go deeper than its founder's love of float trips downriver. Hidden in the Ozark Mountains is a history of the retail revolution told from the ground up. So it is that we venture back to the land to see how Sam Walton built the world's biggest corporation in a little town called Bentonville.

The money that Walton used to seed his retail business came from the U.S. agricultural heartland, but not because he or his family were farmers. Sam Walton's father, Thomas, had tried his hand at farming in Oklahoma in the 1910s, but when crop prices plummeted in the years immediately after World War I, he moved to Springfield, Missouri, where his stepbrother Jesse had a mortgage company. He brought along his wife, Nancy, and two boys, five-year-old Sam and baby Bud. When the Depression set in, Thomas went out across Missouri, repossessing the homes of farmers who could not pay their debts. It was a sad time as a series of droughts hit farmers in Missouri hard and ruined families farther west who became literally buried in the Dust Bowl, one of the worst man-made natural disasters in American history. In his biography, Sam Walton recalled traveling with his father for work during these years and seeing "wonderful people whose family had owned the land forever" being stripped of everything they had.[7]

When Walton, years later, began scouring the land for choice properties for his stores, he was in many ways following in his father's footsteps. Both were effectively in the real estate business. They looked at the physical landscape with an appraiser's eye. Despite his trucker hat—one he wore even when forced to accept fancy awards in tuxedos—Sam Walton, like his father before him, never made his living trucking or tractoring (though Sam certainly paid others to truck Walmart wares and put farmers' produce on his shelves). Farmland in America's heartland was an asset that could become part of an investment portfolio, not a place the Waltons used to raise food from the ground. Historian Bethany Moreton put it well: "Though his father was in fact repossessing those farms for the family mortgage company, Walton's childhood somehow converges with that of the busted farmer, not the moneylender."[8]

The family business earned enough money to enable Sam Walton to enroll at the University of Missouri in 1936. Four years later he graduated and took his first job at a J. C. Penney store in Des Moines, Iowa.[9] There he would begin to hone his skills as a retail salesman.

Department stores like J. C. Penney were common in American cities by the 1940s. Quaker Rowland H. Macy had been one of the early pioneers of the department store format in New York City right before the American Civil War, and within a few decades Macy's red star—copied from a tattoo Macy had put on his body while working as a Nantucket whaler—had spread across the nation.[10] In 1902 in Wyoming, James Cash Penney—a name exceptionally fitting for someone who would make a career out of taking customer coin and making cash registers ring—started his own department chain, which became a fixture of the dry goods market in the American West before becoming a nationally recognized brand.[11]

J. C. Penney and Macy's were just part of a larger retail revolution that swept the nation at the turn of the century. In 1879, Frank W. Woolworth created a variety store that would become the model for many five-and-dime store chains, including Walton's original store concept, in the twentieth century.[12] In 1897, S. S. Kresge opened a similar chain store in Memphis that would one day become Kmart.[13]

In these same years, mail-order retailing also became popular. Aaron Montgomery Ward launched the first mail-order delivery business from Chicago, attracting competitors like Richard Sears and Alvah Roebuck, who would utilize the Windy City's rail lines to ship goods to customers in rural America.[14] Montgomery Ward initially saw members of the Grange, an organization of farmers devoted to populist reform that started in the American South, as some of its most loyal customers. Grangers had grown frustrated with merchant middlemen who they believed were gouging working-class folks in farm country. They wanted to have direct access to cheaper goods produced by mass marketing companies. As a result, southern agrarian reformers helped push for legislation that helped mail-order companies serve rural markets across the country. For example, Georgia populist Tom Watson and North Carolina congressman Marion Butler fought to get a bill passed in Congress that created what was known as Rural Free Delivery (RFD) postal service. First adopted in 1896, RFD tasked the U.S. Postal Service with opening new routes of mail delivery to rural communities. Previously, people living outside urban areas had to travel to town centers to get their mail, and sometimes this meant farmers went more than a week before they found time to get to the post office. Now, with RFD, the postal service brought mail right to customers' doors in the countryside. This, coupled with the introduction of the U.S. parcel post service in 1913, helped mail-order companies like Sears and Montgomery Ward make big profits as country capitalists.[15]

The grocery business was also changing at the end of the nineteenth century and the beginning of the Progressive Era. The Great Atlantic & Pacific Tea Co., or A&P, led the way in the late 1800s in New York City, offering low prices for goods such as tea, coffee, sugar, and other foodstuffs by leveraging their big buying power, managing their own inventory warehouses, and adopting advanced accounting techniques.[16] Others soon followed. Memphis-based Piggly Wiggly, founded in 1916, developed the self-service format that became the standard in the grocery industry, and in the 1930s an employee working for the Kroger Grocery and Baking Company named Michael Cullen took the self-service idea and added another ingredient: large-format, warehouse-like stores built on dirt-cheap property offering an exceptionally wide variety of goods at bare-bones prices. Though Cullen's vision was originally rejected by higher-ups at Kroger, he went out on his own and opened the King Kullen grocery store in Queens, New York, in the 1930s. Thus was born the nation's first supermarket.[17]

In the language of economics, what all these firms were doing was rationalizing supply chains. By cutting out middlemen, working directly with manufacturers, and pursuing a profit model that was about making big-volume sales of low-priced goods, each of these firms acted like blood thinners in the commercial arteries of the American economy.

But not everyone thought these department stores, mail-order houses, supermarkets, and variety chains were a good thing. By the 1920s and 1930s, local merchants fought for their lives, pushing for local and federal legislation that might thwart the spread of these businesses, which represented an existential threat to smaller operators.[18] But despite passage of legislation and small local victories, by 1940 the chains, supermarkets, and mail-order houses used the promise of cheap goods to outlast their opponents. In the hard times of the Great Depression, the promise of low prices made sense for many families struggling to survive. The chains were here to stay.[19]

When Sam Walton took his job in Des Moines, J. C. Penney owned over 1,500 stores in the United States and was the second-largest general merchandise retailer in the country. Only Sears, a company Walton briefly had considered working for before deciding on Penney, did more retail business in this market.[20] He had landed "with the Cadillac of the industry," as he put it, and he was immediately hooked on the salesman gig. "I know this for sure," he said years later of his Penney experience, "I loved retail from the very beginning."[21]

But Walton's love affair, like those of many other young men, was put on hold when the nation went to war. Walton, an ROTC (Reserve Officers' Training Corps) man, immediately reported to the local recruitment office, where he thought he would get the nod to head overseas to the frontlines of battle. But that was not to be. A heart murmur kept him grounded at home, and he was told he would have to wait for further assignment. "This kind of got me down in the dumps," he later recalled, but things soon improved when he ran into Helen Robson at a bowling alley in Claremore, Oklahoma, a place he'd ventured to in his dejected state in hopes of finding a new job and a fresh start. Within a year he would ask Helen to marry him.[22]

Sam Walton's marriage to Helen proved incredibly important for his financial future. Helen's father, Leland Stanford Robson—named after the railroad tycoon who founded Stanford University and whom Robson's parents had met, fittingly, on a train —was a rich man, and in the 1940s he offered substantial financial support to Sam Walton when the young newlywed opened his first five-and-dime in Newport, Arkansas, in

1945. Robson owned a ranch, practiced law, and ran a bank in Claremore, Oklahoma, but the real source of his wealth came from below ground.[23]

In the 1930s, Robson had thought he might go under. His law practice was not bringing in enough money to pay the bills, and his ranch, like many Oklahoma ranches in the Great Depression, was certainly not a moneymaker. It was ulcer inducing. One of his kids remembers that he was reduced to drinking milkshakes and raw eggs to keep down the acid that was corroding his stomach. But Robson's salvation lay buried in the land under his ranch: a monstrous lode of coal. Between 1937 and 1946, according to Robson's son Frank, the Sinclair Coal Company was able to extract dozens of railroad cars of black gold from the property daily.[24] And this was just the start of Robson's fossil-fueled financial boom. In the years ahead, Robson got heavily involved with Oklahoma oilmen, who used a bank he owned to channel money from the statewide oil boom then underway.[25]

Robson shared his riches with his new son-in-law when he decided to open a small 5,000-square-foot store in Newport, Arkansas, in 1945.[26] The $20,000 loan went toward the purchase of a store that would be part of the Ben Franklin variety store franchise owned by the Butler Brothers Company of Chicago and St. Louis. This money, disembodied from the fossil fuel commodities that generated it, seeded the first store of the Walmart empire thousands of miles away.[27]

Sam Walton points to Helen Robson when explaining why he chose to open his first 5,000-square-foot store in Newport, a town of 7,000 people.[28] "Man," he said, "I was all set to become a big-city department store owner." But Helen "laid down the law." She "loved the outdoors" and allegedly told him, "I'll go with you any place you want so long as you don't ask me to live in a big city."[29] So they opened shop in this small town, and business took off. The first year, Walton said, the store did $105,000 in business, and within less than three years, he was able to pay back his father-in-law's loan.[30] Walton was right: he was an exceptional salesman, and he beat out his competition by paying close attention to price, offering promotional deals where he would look to sell more items at smaller per-unit margins than his competitors did.[31]

The good times in Newport, however, did not last. Walton's landlord, eyeing the profits Walton was pulling in, wanted to buy out the Ben Franklin store and refused to renew Walton's lease.[32] "I felt sick to my stomach,"

Walton remembered. The year was 1950 and he now had four kids—Rob, John, James, and Alice—to feed.[33] He now had to start fresh.

That was when he set his eyes on Bentonville, a town even smaller than Newport, nestled near the Ozark Mountains of northwestern Arkansas.[34] Again the Waltons' interest in the outdoors played a big role in the decision: "I wanted to get closer to good quail hunting," he explained, "and with Oklahoma, Kansas, Arkansas, and Missouri all coming together right there, it gave me easy access to four quail seasons in four states."[35] And again with the financial support of his father-in-law, in 1950 he opened a new 4,000-square-foot Ben Franklin franchise, calling it Walton's Five and Dime. Soon thereafter, Walton was back in the black.[36]

Beyond the quail it contained, Bentonville proved an ideal ecosystem from which to build the Walmart empire. Perhaps the most important element was the Ozarks' topography, which forced Walton to take to the air when planning store expansion. By 1957, Walton had opened several other stores, still under Butler Brothers franchise, in Arkansas and neighboring Missouri and Tennessee, and to get to these stores, he often had to take winding roads through mountainous terrain. That got old real fast, and it pushed him to get his pilot's license so he could jump up and over the mountains to scout potential sites for new stores.[37]

Bentonville's unique geography had forced Walton to do something few other store owners were doing at the time. "We were probably ten years ahead of most other retailers in scouting locations from the air," he said, adding, "I guarantee you not many principals of retailing companies were flying around sideways studying development patterns." "From up in the air we could check out traffic flows, see which ways cities and towns were growing and evaluate the location of the competition—if there was any. Then we would develop our real estate strategy for the market."[38] To get around hills hindering his path, Walton went skyward, where he began to see the retail landscape like no one else had before. "Once I took to the air, I caught store fever," he said.[39] Seeing the landscape from the clouds inspired a vision of conquest. He wanted more and more, and over the next two decades the company's "ultimate real estate strategy," as one Walmart executive put it, "was to saturate a state" until there was nowhere else to put a Walmart store.[40]

In addition to spurring Walton's aerial siting strategy, the Ozarks also served another important purpose: they insulated Walton from other retailers.[41] Sam Walton liked to say of his Walmart headquarters, built in Bentonville in 1969, that sitting in his office, he could "hide back there in the hills" out of his competitors' sight.[42] And there was a lot of truth to this. The mountainous terrain and rural nature of Bentonville and its surroundings inhibited the expansion of road infrastructure—a word the "aw-shucks," folksy billionaire Sam Walton said he couldn't even pronounce during a 1990 congressional hearing on federal funding for Arkansas internal improvements.[43] Few big players wanted to venture into this world of winding roads.

Putting down pavement became essential to Walmart growth, but that was often after Walton had decided where he wanted to site his stores. Infrastructure development need not necessarily precede the development of a store in a rural area. Walton explained this master strategy in his biography: "I think our main real estate effort should be directed at getting out in front of expansion and letting the population build out to us."[44] He came to see that there was a lot of money to be made not by placing stores in markets that were already established but by becoming *the* market in places where there was none. This proved a smart business strategy, but it ultimately led to sprawl-like development that threatened the very rurality that had made Walmart great.

There was one more thing the Ozark Mountains did that helped Walton get his business off the ground in Bentonville. In the 1930s and in the post–World War II era, the Army Corps of Engineers eyed the hilly geography of Walmart's hometown as an ideal spot to build several dams. It was a win-win situation in the minds of New Deal planners. Building the dams would create thousands of jobs, and the lakes these dams made would attract recreation dollars to the region. According to historian Bethany Moreton, by the 1960s "the Ozarks had become one of the country's few four-season retirement destinations, a back-to-the-land Florida without the bikinis." Over the next several decades, "weary Midwesterners forwarded their Social Security checks to their new addresses near the region's artificial lakes, courtesy of the Army Corps of Engineers."[45] Like Coca-Cola and Delta, Walmart was born in a region filled with politicians who expressed love for antiunion, laissez-faire, free-market economics but who benefited immensely from New Deal programs and Cold War federal investments. Folks poured into Walton's stores seeking flip-flops, sunscreen, bathing suits, and more. In short, this man-made-lake landscape, made possible in part by the hills and valleys of the region and the interventions of the federal state, helped bring the customers that would help get Walmart off the ground.

By 1962, having had great success with his Benjamin Franklin franchise stores, Walton was ready to go big—really big. That year he opened the first Walton Family Center Store in the 1,500-person town of St. Robert, Missouri. It was a massive retail outlet that ultimately spanned 20,000 square feet—four times the size of his first store in Newport.[46] It was an unprecedented move. Retailers like Memphis-born Kmart and Minneapolis's Target (an offshoot of the regional Dayton's department store chain) had caught the box-store bug in 1962, building their own huge outlets designed to move large amounts of low-priced goods through self-serve, warehouse-size "discount stores."[47] But the big players remained in the city, believing small markets in rural America were not worth the investment.[48] Fifteen hundred people? How on earth could such a town generate enough cash flow to make a big-box store profitable? So said the naysayers, but Walton discovered in Saint Robert that what some dubbed impossible was in fact very profitable. In the early years, he reported $2 million in annual sales, an astounding sum that dwarfed the $200,000 to $300,000 annual earnings Walton's smaller variety stores pulled in.[49]

Seeing the potential for enormous profits, Walton went forthwith to the Butler Brothers' Chicago headquarters to plead his case. He told his bosses he wanted to open more large-format, discount stores in rural America. There was "much, much more business out there in small-town America than anybody, including me, had ever dreamed of," he told them. All they had to do was follow him down these rural roads to riches.[50]

But they turned down his offer. If he was going to make his vision a reality, he was going to have to go it on his own. On July 2, 1962, he opened his first Walmart discount store (then branded Wal-Mart, with a hyphen and capital *M*), in Rogers, Arkansas. Rogers only had about 6,000 citizens, a total population that was less than 40 percent of the total square footage of Walton's store. Within a year, the big store in the tiny town had done $700,000 in sales.[51]

Once Walton got his start, he purchased cheap real estate and opened big-box stores at a frenzied clip, often by repurposing existing buildings like an old Coca-Cola bottling plant in Morrilton, Arkansas.[52] Walton sought to save costs at every turn. The bottling plant turned Walmart outlet—made available as the Atlanta soft drink company slowly consolidated its once widely distributed bottling network—was barely touched up, with "pipes sticking out of the floor and no air conditioning."[53] Walmart rented rather than owned most of these buildings, outsourcing the costs of insurance coverage and real estate taxes to third-party owners. In 1970, the total rent Walmart paid for its thirty-two stores was $500,000, or 97 cents per square foot. Leasing buildings few people wanted in rural America was one of the ways Walton could keep costs down—which was essential if he wanted to offer "everyday low prices" to his customers.[54]

Another way he saved money was by tapping into a cheap labor market, made possible by farmer flight from the fields of Middle America. Agribusiness, supported by large federal subsidies, new machines, synthetic fertilizers, and hybrid seeds, had found a way after World War II to plow and plant huge tracts of land with fewer and fewer farmhands. Growers' relationship with the natural environment was changing. Land that had once supported modest family farms, even in the mountainous and isolated Ozarks, was swiftly becoming the domain of consolidated agricultural concerns. People pushed off the land, including many poor sharecroppers, were desperate for jobs, and the trucking gigs and sales opportunities Walmart offered became, for some, the only way to put bread on the table. As historian Bethany Moreton put it, "The agricultural revolution of the early postwar era was in full swing, depopulating Arkansas farms, and putting tens of thousands of white women and men in search of their first real paychecks."[55] It helped that these people, challenged by the natural elements in America's sun-baked farm country, were primed for hard work.[56] They were the ideal labor force for a firm that looked to drive its workers to accept tough hours for little pay.

Walmart workers, whether women or men, found few protections from southern governments, which proved more interested in attracting businesses to their states than in helping their citizens who were struggling to deal with the sea change in American agriculture. In 1944, the Arkansas legislature voted in favor of one of the country's first antiunion "right-to-work" constitutional amendments, setting a precedent for other southern states, which soon followed suit with similar amendments or laws.[57] For his part, Walton hated unions, and he fought hard against those who wanted to organize Walmart workers.[58] There would be no union presence within Walmart stores for years to come. Displaced farmers in the Ozarks and beyond would work for Walmart on Walmart's terms.

Facing little pressure from regulators or labor groups, Walton paid many of his workers poverty wages. In 1965, Walton ignored new legislation that set the minimum wage at $1.15 an hour, paying some of his employees less than half that amount. Ultimately, the federal government sued Walmart and forced Walton to release back pay to his workers, but in the years ahead Walmart managers continued to find ways to underpay employees, including by forcing employees to work overtime without compensation.[59]

Instead of using high wages to engender worker loyalty to his firm, Walton did something else. First, in 1971 he created a profit-sharing program that allowed workers to acquire company stock at discount rates after the firm went public in 1970. Walton believed that encouraging his employees to become Walmart stockholders would help defuse any grumbling about low wages.[60] In addition to this program, Walton also sought to create a corporate culture that served as a kind of salve for young men being wrenched from the land. He created a gendered workspace in his stores: women served in low-pay sales and service positions, while men took on managerial responsibilities. He sent many male managers to nearby Christian universities, where they were indoctrinated into evangelical traditions of servanthood as they learned business basics. While Walton made billions, his managers earned modest salaries, but they came to accept the financial shortcomings in return for the intangible benefits associated with being atop a patriarchic hierarchy within a new corporate "family" held out as a proxy of the old family farm.[61]

Walmart's long-haul trucking network—the largest in the world by the end of the century—proved attractive to displaced farmers who desperately wanted independence and freedom but could not find them in the corporate agribusiness landscape of the postwar era. As New Deal–era farm policies channeled millions of dollars into the pockets of large landowners, machines replaced rural workers, who were forced to trade their tractors for trucks to put bread on the table. Many of these men became deeply resentful of the government, which they saw as being the source of the agrarian problem rather than the solution, and adopted a hardcore allegiance to free-market, antiunion politics. This was extremely beneficial for Walmart, because it meant truckers avoided the minimal opportunities for collective bargaining that still existed in Arkansas and instead engaged in cutthroat competition for business that drove down Walmart's shipping costs.[62]

For these truckers, the open road became a symbol of independence from government control, even if the road these truckers were on was entirely a government creation. After all, the Interstate Highway Act of 1956 funneled nearly $25 billion in taxpayer dollars into the construction of the infrastructure that made Walmart's world possible. Without that investment, Walton would never have been able to realize his discount dream in small-town America. Though he, like the truckers who worked for him, maintained a staunch commitment to free-market principles, it was in fact the government that paid for the federal highways that allowed Walmart to grow.[63]

Access to cheap land and cheap labor were critical ingredients that helped Walmart spread—"like a weed," as Sam Walton put it—in the disturbed soil of rural America in the 1970s and 1980s.[64] But the firm needed capital if it was going to break out from Bentonville. By 1970, Walton had only thirty-two stores up and running. To put that number in perspective, Kmart had already built 200 discount stores in metropolitan areas across the country by 1969.[65] Even though Walton had low overhead, early expansion had already put him deeply in debt. He owed local bankers, and the only way he could pay them off was by bringing in money from outside sources.[66]

That was when Little Rock–based investment firm Stephens Inc. got involved. In 1970, Stephens Inc. agreed to handle Walmart's initial public offering, which resulted in the sale of 300,000 shares at $16.50. The stock issue was modest, with only 800 buyers willing to take a risk on the small Arkansas firm, but it was enough to get Walton out of the red and to help him finance the next phase of store expansion.[67]

The pace of growth immediately picked up, and Walton expanded to seventy-eight stores by 1974. That year Walton, now in his mid-fifties, made a premature decision to retire from the company, briefly handing over the reins to Vice President Ron Mayer, but in less than two years, clearly too driven to give up control of the company he had founded, Walton returned, and when he did, he hastened the pace of company growth.[68]

Pushed by computer-savvy CFO David Glass, Walton invested in an automated distribution system in Searcy, Arkansas, that soon became the envy of the industry. Utilizing UPC scanners (first introduced in 1974), sophisticated computer networks, and cross-docking techniques that allowed goods to seamlessly flow onto loading docks and out in trucks on the other side, Walmart radically reduced the time inventory sat in warehouses. All this helped the company achieve the price point that helped cash-strapped Americans in Middle America purchase the consumer goods they wanted.[69]

Year after year, Walmart followed the same formula: building state-of-the-art distribution centers in rural parts of the American South and the Midwest and then filling in stores around those centers. In 1981, the company punched deep into the Deep South, acquiring the Nashville-based Big K chain for $13 million. It then looked west, opening a distribution center in Mount Pleasant, Iowa, in 1984. The company also diversified, expanding into the grocery business, especially through new Sam's Clubs, warehouse-style grocery outlets patterned on California-based competitor Costco. All the while, the company sank lots of capital into communications technology. By the end of the 1980s, the company had launched a satellite system that would streamline information sharing between all its stores scattered across the country. Soon Retail Link went online, with Walmart pioneering cutting-edge electronic data exchange practices with its suppliers that enabled swift restocking of sold items. These tools proved key in making Walmart the undisputed leader of the retail industry by 1991, a year after Walton's retirement and a year before his death.[70]

Walton pointed to geography as the key factor that helped Walmart develop its winning way. "We were forced to be ahead of our time in distribution and in communication," he said, "because our stores were sitting out there in tiny little towns and we had to stay in touch and keep them supplied." Rurality, something many believed hindered capital flow, proved essential in stimulating the development of sophisticated distribution systems that radically accelerated the pace of commercial exchange. The financial capital that seeded this revolution may have come from outside the American South, but Walton's system was nevertheless an Ozark creation, adapted to the environment in which it was born.[71]

Yet if rurality helped give birth to Walton's empire, it also threatened its future. Beyond the American South, in hills and valleys known for their natural beauty, cries of resistance to Walmart's retail revolution could be heard. In the waning years of the twentieth century, the epicenter of the fight was a small town called Williston, nestled in the Green Mountains of Vermont.

Chapter Six

SOME THINGS JUST DON'T FIT IN WILLISTON

The battle intensified in 1991 when a local developer, Taft Corner Associates, proposed siting a Walmart store in Williston, Vermont, population circa 4,800. Located just outside Burlington, the town had long attracted people fleeing New England cities in search of something more rural; Malcolm Gladwell, writing for the *Washington Post*, dubbed them "ecotopians."[1] This included Jerry Greenfield and Ben Cohen of Ben and Jerry's fame, business partners who shared a worldview perhaps best encapsulated by the catchy names of the firm's ice creams, like Cherry Garcia and Wake and No Bake. Ben and Jerry, who had started their business in Burlington back in the 1970s, were going back to the countryside in search of serenity, as farmers in Middle America continued to leave the rural world in search of city jobs.[2]

Walmart had no doubt wanted to come to Williston because it had all the environmental ingredients the firm liked. An isolated mountain community of just a few thousand people, an Ozarks in the North, this was just the kind of place, devoid of discount competitors, where Walmart could make millions.

Yet if it wanted to do that, the firm was going to have to change the very rural natural environment that had attracted it—and Ben and Jerry—to this place. After all, digital communication systems and outer-space satellite technology, could not, in and of themselves, move Walmart's goods to market. The firm had to touch the ground—or the pavement, as it were—to get the goods it sold to their stores, and in towns like Williston, the blacktop simply wasn't adequate for this retail giant. Rural roads, the on-the-ground arteries of Walmart's commercial network, would have to be widened to allow Walmart's trucks and out-of-town customers access to this market. This was a change many Williston residents would resist.

The Williston Citizens for Responsible Growth (WCRG) led the charge against Walmart. In 1988 Ben and Jerry's CEO and Williston resident Fred "Chico" Lager formed the group with like-minded citizens—including Jerry Greenfield and his wife—in order to combat a series of commercial development projects that threatened to turn this little Vermont town into "any other place, U.S.A.," as Burlington mayor Peter Clavelle put it.[3] By the time local developer Jeffrey Davis decided to propose building a Walmart on the Taft Associates commercial site, members of the WCRG had already successfully blocked an attempt by a New York firm to build a mall in the town's central corridor—which had just one yellow blinking light at its central intersection. The WCRG was organized, experienced, and ready to fight.[4]

And they had a powerful weapon at their disposal: Act 250, a law passed by the Vermont legislature in 1970 designed to prevent sprawl-like development that could sully the state's natural beauty. Passed in response to concerns about a proposed ski resort, the law required developers to conduct extensive environmental assessments to ensure that a project would not cause any undue harm to Vermont's ecosystem. It was a unique piece of legislation, a law Walmart had not encountered in other states. In 1991, the firm had failed to site a single Walmart in Vermont. This was the final frontier of Walmart's American conquest.[5]

Though Walmart executives saw an ecological environment ripe for their brand of big-box development when they initially decided to come to the Green Mountain State, it quickly became clear that they were headed into a political ecology very different from the one they enjoyed in their home state of Arkansas. Though Vermont was one of the most rural and whitest states in America by the end of the twentieth century, in the 1990s it was also becoming more Democratic. Eco-conscious liberals in big cities like New York and Boston had begun an exodus to the Vermont countryside in the 1960s and 1970s, drawn to the state's relative rurality and to laws like Act 250, which signaled to newcomers that politicians would fight to protect the mountain environment there. By contrast, Arkansas's political elite, like those in many other southern states, had long promoted lax environmental regulations as a way of selling their state to entrepreneurs and investors.[6]

Of course, many conservatives also lived in Vermont, and national politics too had swung decidedly to the right in the wake of the Reagan Revolution. Bill Clinton, former Democratic governor of Arkansas, was in the White House by the 1990s, having won a national following, including in Vermont, by rebranding the Democratic Party as centered on reducing the size of government and reforming welfare, policy prescriptions that had long been popular in the American South among conservatives who felt that federal efforts to address structural inequalities brought on by Jim Crow segregation were unwarranted. These policies found favor among white suburbanites in other parts of the country who were drawn to the policy cocktail of consumer choice and individualism that these "New Democrats" served up. Clinton adopted the kind of race-neutral language that cast suburban families' desire for limited government intervention in their lives and good schools as merely the meritocratic strivings of middle-class Americans who wanted to enjoy the fruits of their hard work.[7]

It seemed that many middle-class suburban voters across the country, whether in New England or the American South, were converging on similar political values forged in the wake of white flight from urban areas. The fact that Hillary Clinton served on Walmart's board in the 1980s was also another indication that Walmart's antiunionism and free-market evangelism had become more acceptable in a Democratic Party that once had put blue-collar AFL-CIO members at the center of party politics.[8]

But even though a common political discourse that emphasized middle-class voters' right to consume goods in a free market unhindered by government red tape had swept the nation, there were still real differences between Vermont voters and Arkansas citizens. Most notably in Vermont, a growing coalition of voters expected their state government to play an aggressive role in preventing the degradation of the natural environment. Rurality, or at least a type of rurality that appealed to affluent residents, was something the law was supposed to preserve. Walmart was set to face fierce resistance from these citizens as it pushed to site its store in Vermont.

At first, it seemed Walmart's odds looked good in Williston. Jeffrey Davis had already navigated the Act 250 approval process in the 1980s and had an umbrella construction permit that ostensibly gave him the right to build a Walmart in Williston. But in the early 1990s, the WCRG sought to make the case to the state's regulators that they had not considered the huge impact a Walmart would have on their community when granting Davis's permit. How could the commission grant Act 250 approval before knowing exactly what Davis planned to build on his site?[9]

The WCRG decided that their best tactic was to focus on roads and traffic, covered under provisions of Act 250. During the original permitting process, the local environmental commission had clearly said the Taft Associates development "will be community or county oriented," and it went on to state, "The Interstate has not been represented as a necessary link." This obviously meant, Chico Lager explained, that a Walmart was not something regulators had envisioned in Williston.[10]

By design, Walmart was a company that made its money by attracting customers from dozens of miles away from a store site. It had to. To achieve huge volume sales of low-priced goods, the firm could not depend on the few thousand folks who lived in Williston. If Davis built his Walmart, it was going to bring a flood of out-of-town traffic to Williston.

The WCRG thought they had Walmart in a tough spot. A traffic study in 1992 made clear that millions of dollars in roadway improvements would have to be completed before a Walmart could be sited in town, and much of that money was going to have to come from taxpayer dollars. The municipality and the state would probably be tapped for more than $15 million in infrastructure costs. With local and state governments facing tight budgets, there was little chance roads could be widened fast enough to warrant Walmart's entry.[11]

WCRG, partnering with the city of Burlington and Vermont's Agency of Natural Resources, took its case to the state's Environmental Board, which required the district commission to reevaluate Davis's permit per WCRG traffic concerns. Taft Associates naturally appealed the decision to the Vermont Supreme Court, which then tasked the Environmental Board with settling the affair. In 1994, three years after WCRG began its wrangling over Walmart, the Environmental Board ruled that Jeffrey Davis's project could go forward, but not before an estimated $15 million in road infrastructure was completed.[12]

The WCRG thought it had won.[13] These costs were simply too high. Some people thought there was no way the town could make these adjustments anytime soon. Ironically, Williston's rural roads, the thing that had created the discount retail desert that brought Walmart here in the first place, had temporarily forestalled Walmart's expansion into its final frontier. "Some things just don't fit in Williston," a WCRG protest poster read.[14]

But the battle continued. In the spring of 1995, Jeffrey Davis appealed the 1994 decision and got a serious gift from the Environmental Board: a radical reduction in total outlays required before the Walmart project could go forward. Instead of the $15 million in infrastructure improvements initially projected, the board required that only $3 million worth of work be completed before Walmart could break ground. This helped get the Arkansas firm one step closer to landing its first Vermont store.[15] By summer, WCRG had received more bad news: Davis said he was going to front much of the money for roadway improvements himself. He was tired of waiting; he would go it alone. Davis did not disclose just how much money he was going to spend to get the job done, but proposed construction plans included widening on and off ramps on I-89, expanding roadways that cut through town, and making major intersection improvements.[16]

The *Burlington Free Press* ran a front-page story under the headline "Wal-Mart to Pave Own Way," but this was not entirely true. Davis made this clear in his statement to the press, saying, "It's our money. It's not Wal-Mart's or anybody else." The truth was that local dollars were being spent by a local developer to remake a local landscape, even if many locals didn't like it.[17]

WCRG would not give up: in the fall of 1995 it appealed the Environmental Board's decision to the Vermont Supreme Court.[18] But the court ultimately decided in favor of Davis and Walmart. The Environmental Board's decision would be the final word on the matter. "As we understand it," a happy Jeffrey Davis told the press, "this is the end of the road." It was a fitting statement for this developer turned road maker.[19]

After that, things got nasty. Eco-saboteurs, stealing a page from environmental radical Edward Abbey's classic *The Monkey Wrench Gang*, took matters into their own hands, sneaking onto the Walmart construction site after hours and dumping a mix of "rice and cornmeal into fuel tanks and hydraulic systems" of earthmoving equipment. It was an expensive act of defiance, one that set a locally contracted construction firm back a few thousand dollars. For Walmart, this was the advantage of operating at arm's length. Because the company typically leased its stores from local developers, the company avoided taking on certain risks and financial liabilities associated with store construction. In this case, monkeywrenching carried out by eco-crusaders eager to hurt an out-of-state firm ended up hitting the pocketbooks of people closer to home.[20]

And it did little to stop Walmart from coming to town. By January 1997, roads had been widened, the Walmart store had been built, and Williston had been changed.[21]

Not everyone thought this was a bad thing. Even though a local poll showed that roughly 70 percent of Chittenden County residents opposed allowing Walmart into Vermont, the *Burlington Free Press* interviewed many citizens who said they welcomed the low prices Walmart would bring to the community.[22] A mother of five, a grocery clerk, an office cleaner from Burlington, a Walmart applicant from nearby St. George: these were just some of the people who wanted to see the big-box giant come to town.[23] These voices reflected the way the Walmart debate was refracted through the prism of class. For many working-class people in Vermont, Walmart's low prices offered relief at a time when American wages were stagnating and the cost of living was rising.[24]

But for those concerned about the environment, Walmart's cheap goods came with a high price tag. Pavement, after all, did not just bring a flood of traffic to town. It also increased the flow of stormwater into drains and the release of harmful air pollutants from idling autos. The Williston Planning Commission had talked about stormwater concerns in 1987 when Taft Associates was still in the design stages of development, noting that much of the runoff from the proposed site would dump into nearby Muddy Brook.[25] But that was before Jeffrey Davis decided to put a Walmart on his site. As the Environmental Board explained in its 1992 review of Davis's permit, the new project proposal included plans for much more paving. Over 50 percent of Davis's property would soon be covered by buildings, huge parking lots (capable of accommodating 4,900 parked cars), and roads, up from 30 percent in the original proposal, which raised the "potential for impacts not previously considered."[26]

The Environmental Board was right to be concerned about Walmart's effects on stormwater discharges. Just a few years after Williston's Walmart opened its doors, the EPA toured Walmart construction sites around the country and "detected a pattern of failures to comply with the requirements of applicable permits for the discharge of storm water from these construction sites."[27] "Inadequate silt fences," faulty retention ponds, and "unprotected slopes" were just some of the problems cited in Delaware, Utah, Colorado, and six other states.[28] In 2001, Walmart agreed to a $1 million settlement with the EPA, but three years later, the environmental agency found that Walmart was still not in compliance with basic stormwater rules.[29] The EPA negotiated another settlement with Walmart for $3.1 million, "the largest civil penalty ever paid for violations of storm water regulations."[30]

In Williston, Walmart contributed directly to water problems. Muddy Brook became an impaired stream, and environmental engineers specifically pointed to Walmart's parking lot as one of the key culprits contributing to tributary erosion and habitat loss.[31] "Stormwater runoff" was one factor "causing a decline in biotic integrity," a 2009 watershed report read, noting that the urbanization in the Taft Corners area was exacerbating problems. "Without steps to address the watershed," the report concluded, "habitat conditions will continue to decline and will be less likely to support a reference biotic community in the future."[32]

And then there were the issues with air quality. Environmental regulators had required Davis to get an air quality permit in 1987 because they "determined that the installation of more than 999 parking spaces . . . may generate nonpoint source air pollution." It was essential, the commission noted then, to ensure "that good traffic flows are maintained so that excessive emissions do not result from idling autos."[33]

This did not come to pass. In part because Davis only made minimal changes to roadways, not the investment of over $15 million that experts had suggested in the early 1990s, intersections and turning lanes quickly became clogged, with traffic often stalling all the way up the exit ramp on I-89.[34] It was a nightmare for many commuters, especially around rush hour.[35] Ultimately, the Vermont Agency of Transportation had to intervene, throwing millions of dollars at infrastructure expansion in Williston to try to fix the problem. So much for Walmart paving its way. In the end, it took state dollars to deal with the inevitable traffic problems this big-box store helped bring to this small town.[36]

But in the meantime, those backed-up cars belched out pollutants, precisely as the district environmental commission predicted, though just how much this changed the local air in Williston is not precisely known, in part because the EPA never put an ambient air quality monitoring system in the town. With limited resources and budget, the EPA installs ambient air quality devices only in bigger cities, and the closest one to Williston was in Burlington, over six miles away. Here was an advantage Walmart gained by siting its stores in tiny hamlets: these towns were often devoid of EPA instruments that might track the fine-grained ecological impact of a particular store.[37]

But air pollution was in many ways out of sight and out of mind. What really bothered those who had fought so hard against Walmart was the new look and feel of the town. After Walmart came a flood of other big-box stores—Home Depot, Best Buy, Toys R Us, and more. It was as if Walmart had opened the door for a whole slew of new development. As many saw it, a natural beauty and an ecological aesthetic that defied precise valuation had been damaged and would never be fully restored. In the words of environmental historian Jack Temple Kirby, these were "rural worlds lost."[38]

Of course, Williston was not alone. In the 1980s and 1990s, dozens of small-town residents across the country fought Walmart on the grounds that the firm destroyed a certain rural aesthetic. In these battles—and there were many—one of the key concerns was that Walmart's entry would lead to the decline of local businesses and Main Street establishments. Beginning in 1988, Kenneth Stone, a professor and economist at Iowa State University, provided the statistics supporting this claim: he published studies showing that Walmart hurt many local shops and stores over the long run (typically after five years). Digesting nearly ten years of data that looked at small towns in Iowa, Stone concluded in 1997 "that some small towns lose up to 47 percent of their retail trade after 10 years of Wal-Mart stores" doing business nearby. Though Stone later (in 2012) produced a study that offered a more nuanced view of Walmart's impact on local businesses and concluded that "Walmart's presence" in small Iowa communities actually "helped stabilize and even expand the local retail sector of most rural Iowa host communities," his original findings emboldened opponents of Walmart stores in the 1990s who turned to Stone's economic studies to argue that the Arkansas discount store was not good for local business.[39]

But while these economic arguments proved popular in Walmart debates, roadways and the environmental change these roadways wrought were often at the center of anti-big-box discussions. In Upstate New York in the early 1990s, the Otsego County Conservation Association fought a Walmart distribution center in the small town of Sharon Springs that routed trucks through rural roads. "We feel it is unfair for us to bear the brunt of the negative impacts we foresee," association president Bonnie Canning-Hofmann told Walmart vice president H. Lee Scott in a 1994 letter, listing among her grievances "increased heavy truck thru-traffic, suburbanized traffic patterns, air pollution noise, disturbance to historic villages and to the rural quality of life we deeply appreciate." Canning-Hofmann wanted Walmart to avoid certain rural roads and proposed a traffic plan that would "route large trucks to the interstate highway system built and intended for their use."[40]

But Scott, writing in response to Canning-Hofmann, explained that the shorter access roads to the highway had "inclines, curves, underpasses, and weight restrictions," making it impossible to route trucks that way. The physics of shipping huge amounts of goods in big tractor trailers meant that Walmart could only take certain routes through rural America. Though Scott said he appreciated her "concerns regarding traffic," Walmart was going to send its trucks down Highway 20—a road Canning-Hofmann described as "one of the most beautiful . . . in New York . . . nestled among bucolic unspoiled hills, dramatic and magnificent view[s], enhanced by the presence of working farms." Scott said she shouldn't worry because the "peak traffic level on this route will not be reached for several years. . . . This should allow all parties time to adjust to the new truck traffic and reevaluate and adjust routes or modify roads if and as needed." It was going to be Walmart's way, even if that way would have to be paved with state dollars down the road.[41]

Walmart did not always win. In Westford, Massachusetts, thousands of residents supported a Stop Wal-Mart Committee campaign in the early

1990s to prevent the Arkansas discounter from coming to town. They used the local environment to their advantage, pointing to pools on the prospective Walmart property that were breeding grounds for frogs in an effort to thwart Walmart's proposed development. Here was another way that roadways and pavement affected the local environment of Walmart sites. Stores designed to increase the flow of out-of-town traffic in a particular community also served to impede the free flow of endangered species migration. But the local developer working with Walmart on the Westford project said this was no matter: he would figure out a way to have a biologist on site who could help frogs get to their breeding grounds—one assumes by directing traffic like a crossing guard for the amphibious. But this never came to pass because Walmart ultimately bowed to the pressure of some 4,000 protestors (nearly a quarter of the town) who simply would not give up their fight.[42]

Steamboat Springs, Colorado; Iowa City, Iowa; Ithaca, New York; Jackson Hole, Wyoming: all these towns, east and west, became sites of anti-Walmart flare-ups in the 1980s and 1990s. And in these battles, ecology, not just economics, entered the lexicon of protest, as some residents in small towns fought to preserve a rural aesthetic that they believed made their towns special.[43] Unlike in Westford, where Walmart's man on the ground said he would pay for a reptilian traffic controller, Walmart could often be crass in its response to protests that focused on ecological preservation. In Ithaca, New York, for example, town members opposing the siting of a Walmart in front of Buttermilk Falls State Park were aghast when a Walmart spokesman glibly said that their little state park was no "Grand Canyon."[44] What was the big fuss all about? he argued. What this spokesman missed was that those knolls and small waterfalls, the little natural wonders of this small town, were prized ecological gems for many local people, even if those gems were not as big as the Grand Canyon.

Despite the protests, Walmart ultimately came to Ithaca, as it did to many other communities that fought the big discounter. Westford was really the exception, not the rule. Between 1998 and 2005, roughly 563 communities protested Walmart store constructions, which represented more than a third of the 1,599 sites proposed during that period. But in the end, Walmart chose to abandon only about 365 projects. Over that eight-year period, the company completed construction on 1,040 stores, bringing its total number of outlets in the United States to well over

3,000. These stores helped make Walmart not only the world's largest retailer but also the largest corporation on the planet.[45]

By opening commercial conduits into the smallest towns, Walmart gained outsized influence on the global economy. Now with thousands of stores attracting millions of rural customers, many of them outside the American South, Walmart could make demands on suppliers who trembled at the thought of losing access to this massive market the company had created. It turned out that the environmental effects of Walmart's empire stretched far beyond the hills of Williston or the frog ponds of Westford.

In 1980 and 1981, Walmart opened its first procurement offices in Hong Kong and Taiwan. The company certainly was not a trailblazer in turning to Asian suppliers to provide cheap consumer goods, but it quickly became a big player in this market. In 1985 Sam Walton announced, with much fanfare, his "Buy American" campaign, pledging to source much of the materials in his stores from domestic companies rather than overseas suppliers.[46] But in 1992, *Dateline* reporters revealed the hollowness of Walton's patriotic appeal, taking viewers on a virtual tour of Walmart stores and showing tags on "Made in America" racks that revealed merchandise coming from China, Bangladesh, and other countries overseas.[47] The exposé also documented horrendous working conditions in sweatshops that supplied the firm in Southeast Asia. The truth was that Walmart, despite its best efforts to bill itself as American born, was a hawker of foreign wares. By 2002, 80 percent of Walmart suppliers were based in China.[48] That year, the company ranked ahead of Russia and the United Kingdom in terms of the trade it did with China.[49]

But Walmart was not just an international company because it sourced goods from overseas; it also built stores beyond U.S. borders. Walmart opened its first retail outlet in China in 1996, a year before anti-Walmart Willistonians lost their fight in Vermont, and in the 1990s and 2000s, Walmart, like many American companies, took advantage of neoliberal trade policies promoted by new organizations like the World Trade Organization to open stores across the globe.[50] The company's first foreign foray was in Mexico in 1991, and then—in 1994, after Clinton signed NAFTA—Canada. Walton saw where Walmart was headed. "I don't know if Wal-Mart can truly maintain our leadership position by just staying in this country," Walton said in 1992. "I think we're going to have to become a more international company in the not-too-distant future."[51] Walmart created an international division and soon opened its first stores in Brazil (1995) and Argentina (1995) and then Indonesia (1996) and China (1996). Europe soon followed, as the firm bought out competitors and opened new stores in Germany (1997) and the United Kingdom (1999).[52]

During these years, the company learned the hard lesson that its retail model did not suit all environments. In Germany, for example, tough zoning laws prohibited the kind of sprawl-like growth that defined Walmart's retail model in the United States. Unable to develop big-box stores in rural areas that would attract customers from dozens of miles away, Walmart decided it simply could not make it work in Germany, and in 2006 it cut its losses and abandoned the country entirely.[53] It had better success in Mexico, a country it entered in 1991, but in Japan Walmart also ran into trouble: living in an island nation where space is limited had shaped consumer habits in ways that put them at odds with Walmart's business model. As geographers Yuko Aoyoma and Guido Schwarz explained, "Lack of storage space provided no incentive for customers [in Japan] to shop at large-volume discount stores." Instead, consumers were "more likely to walk or bicycle to a nearby grocery store or drop by a store after work en route from a nearby train or subway station," making more shopping trips than the typical American. The Japanese real estate market also proved particularly problematic for Walmart, in large part because in the United States the firm had enjoyed dirt-cheap lease prices, mainly by placing their stores in rural areas outside high-dollar urban centers. In Japan, however, land scarcity drove up retail prices, and that hurt Walmart's bottom line.[54]

In 2006, over a decade and a half after the firm began opening stores overseas, international earnings only represented 19.7 percent of sales.[55] Twelve years later, international business remained below 24 percent of consolidated net sales.[56] These statistics made clear that the rural roads in small-town America were still critical arteries in Walmart's corporate body.

Yet these arteries—a widened road in Williston, for example—would never have pulsed if not for the steady heartbeat of Walmart's suppliers thousands of miles away. Because for a working family in Vermont to feel like it was worth it to drive fifty miles to a store in Williston, Walmart had to find ways to drive prices for its goods down as far as it possibly could.

In the industry, it was called the "squeeze," and Walmart was one of the best in the business at doing it. This is how it worked. A company would come to Bentonville, hoping to secure a contract to sell its wares to Walmart, and in that meeting, Walmart, not the supplier, would dictate the terms of the transaction. Walmart would explain what the price would be for a given good, and the manufacturer would be the one who would have to figure out how it was going to meet Walmart's demands. There was no real negotiation. Because Walmart was so big and powerful, they said what cost what.[57]

Chicago-based Lakewood fan company had experienced the squeeze back in the late 1990s. At that time, Lakewood's twenty-inch box fans ran about twenty bucks a pop at Walmart stores, but the discount retailer wanted the company to drastically reduce this price. The only way Lakewood's management team figured they could do that was by outsourcing production to China, where they could pay workers about twenty-five cents per hour, as opposed to the thirteen dollars per hour mandated by Chicago's market. So they opened shop in China in 2000, and a few years later, they sold fans at Walmart at half the price they had originally offered them.[58]

This outsourcing strategy might have made smart economic sense, but from an environmental perspective, it meant that a large percentage of Walmart's goods had to be transported over extremely long distances via cargo ships, resulting in tremendous greenhouse gas emissions. In 2003, Walmart was the top U.S. importer of shipping containers from overseas markets. Atlanta-based Home Depot, a store whose big-box business strategy was very much modeled on that of discount retailers like Walmart, came in second.[59]

All that shipping traffic had a big impact on global climate change. Between 1994 and 2014, container-ship traffic increased by roughly 400 percent, and tankers consumed billions of gallons of heavy fuel oil (HFO), which the *Washington Post* called "nasty stuff," citing the particularly noxious concoction of greenhouse gases and air pollutants this fuel released when burned. And the prospects for improvement were not good: some studies predicted that, without major changes to marine transport, by 2050, total emissions from the shipping industry might more than double.[60]

Of course, greenhouse gas emissions from transport were only part of the problem when it came to Walmart's business in China. After all, another reason suppliers like Lakewood chose China to do business was that there they had to abide by very few environmental regulations. Greenhouse gases and air pollutants belched from energy-inefficient factories overseas. As Walmart admitted, roughly 90 percent of the firm's greenhouse gas emissions came from its supply chain, not its retail stores. When it came to Walmart's ecological footprint, Williston's environmental woes were just the tip of a melting iceberg.[61]

And this reality had Walmart executives worried. In 2004, Walmart's CEO H. Lee Scott admitted that his company, now the largest corporation in the world, was "exposed" on so many fronts. Scott, a Missouri native who first joined the company in 1979, had never really been an environmentalist. In fact, as journalist Edward Humes put it, "the environment was the last thing on his mind" when he became CEO in 2000.[62] But a series of events culminated in Scott committing to "greening" his company in 2004 because he believed it was the only way to protect the firm's bottom line.[63]

Troubles had been festering since the 1992 *Dateline* investigation that had showed Walmart's Bangladeshi suppliers exploiting child laborers. David Glass, Scott's predecessor, had totally botched the *Dateline* interviews, at one point responding to footage of children working in supplier factories by saying to the NBC reporter on set, "You and I might perhaps define 'children' differently." Then, in 1996, labor rights activist Charles Kernaghan testified to Congress that Walmart's Kathie Lee Gifford clothing line was also produced by child laborers working in horrible conditions in Honduran factories. The company said it was taking steps to eliminate these injustices in its supply chain, but clearly there was a deep culture of labor abuse inside the firm. In 2001, for example, a group of women who worked at Walmart filed suit against the firm, charging that the company had systemically paid female employees less than male workers doing the same jobs. Two years later, federal authorities stormed Walmart stores in twenty states, detaining hundreds of illegal immigrants who were employed in these outlets. And at the same time, a confidential internal Walmart report was leaked to the press that made many people gasp. The report revealed that roughly 46 percent of the children of Walmart employees had no health insurance or depended on Medicaid or other public welfare programs to meet their health-care needs.[64]

As all this news was breaking, communities were continuing to battle the environmental (and cultural) degradation caused by the siting of Walmart stores in rural communities. To be sure, these environmental issues were not as big a concern to folks like Scott in 2004, but they added to the distress top executives felt. That year, the EPA issued its historic

$3.1 million fine against the company for its stormwater violations, and the same year the company paid $400,000 in penalties to settle complaints about air pollution caused by Walmart's illegal sale of banned CFC refrigerants. In Mexico, activists expressed outrage when the company announced plans to build one of its stores near the ancient city of Teotihuacán, a sacred UNESCO World Heritage site. And Walmart became a target of environmental activists concerned about the large firm's massive contributions to global warming.[65]

Scott had hard data to confirm his fears that Walmart was facing a potential massive customer rebellion. A confidential study conducted by McKinsey & Company in 2004 revealed that perhaps 8 percent of Walmart customers had already stopped coming to company stores because they were disturbed by the company's labor and/or environmental abuses; 54 percent of those surveyed said that Walmart was behaving too aggressively in its efforts to expand into new markets.[66]

So even though Lee Scott later admitted that he was acting defensively rather than out of some deep concern for the environment, in 2005 the CEO of Walmart made a bold promise to make his company more environmentally friendly.[67] For Scott, it was all about changing the reputation of his firm. He could have chosen to focus on any of the issues outlined above, but many of the labor issues the firm was entangled in seemed intractable. As journalist Edward Humes put it, "The company already faced big lawsuits on labor practices and alleged discrimination; Scott was spending up to an hour a day with the lawyers. But environmental complaints had not reached such a crescendo in 2004, and Scott sensed more room to maneuver on green issues." It also helped that just in this moment an environmental consultant named Jib Ellison came to his office and showed him a new way to look at the ecological concerns the company faced.[68]

Ellison, a one-time whitewater raft guide, had just created a new consulting firm in northern California called Blue Skye that focused on advising businesses on strategies to go green. Ellison had the good fortune in 2004 of having a close friend introduce him to S. Robson "Rob" Walton, Sam Walton's son and the chairman of Walmart's board. Through conversations, Walton became persuaded that Ellison had ideas that might be beneficial to Walmart. He decided to introduce Ellison to Scott, who was at the moment scratching his head trying to figure out how he was going to rebrand the massive corporation he was overseeing.[69]

Ellison made a pitch that really sold Scott on environmental sustainability. He said that Walmart's move toward greening its operation would not just be good for improving its public image; environmental sustainability initiatives could also save the company tremendous amounts of money and thereby increase its profits. To prove his point, Ellison turned his attention to easy adjustments the firm could make, including reducing the size of a cardboard box used to ship a popular toy bear. In this example, Ellison showed that the company would radically reduce the number of shipping containers it needed to transport this item from suppliers to retailers, saving the firm something like $2.4 million.[70]

Scott, hardly a tree-hugger, was sold. Over the next several months, he worked with Ellison to identify other areas where the firm could become more efficient and cut waste. After each consulting round, the evidence became increasingly clear: the firm would save billions of dollars by greening its operation.[71]

On October 24, 2005, to the surprise of many, Lee Scott gave an impassioned speech to company employees, announcing that the firm was henceforth committing itself to using its "size and resources to make this country and this earth an even better place for all of us: customers, Associates, our children, and generations unborn." Noting how company employees had come together to provide aid and relief to people suffering from Hurricane Katrina, which just a few months earlier had devastated coastal areas near New Orleans, he challenged the firm to focus similar energies on solving larger global environmental crises around the world. Before leaving the stage, he promised that Walmart would commit to three major initiatives: powering its operations with 100 percent renewable energy, eliminating all waste in its corporate system, and selling "products that sustain our resources and environment."[72]

These were bold words for a man who just a few months earlier had been virtually silent on the need for corporate environmental stewardship. But over the course of the next decade, the company made major moves, pushing electricity-saving compact fluorescent light bulbs in its stores, radically shrinking product packaging, and improving the energy efficiency of its trucking fleet. Within a decade, the firm boasted that it sourced 26 percent of its electricity from renewable resources and claimed an "87.4 percent improvement" in "fleet efficiency."[73] The company also dramatically increased purchases of organic cotton for its clothing lines—by the 2010s it had become the single largest purchaser of organic cotton in the world—and worked on a series of projects to reduce the greenhouse gas footprint of its dairy suppliers.[74]

Perhaps the most ambitious project was an effort to develop a "sustainability index" that would allow the company to know the entire environmental footprint, from cradle to grave, of every product they sold. Beginning in 2008 the company worked to create a Sustainability Consortium made up of university experts and industry representatives from firms such as Seventh Generation, Procter & Gamble, and SC Johnson, which was tasked with assessing the full life cycle of all Walmart products. As this book went to press, the project was still in existence and known as "the Sustainability Insight System (THESIS) Index." Walmart said in 2022 that it drew on this index to make supply-side decisions but noted that only 1,800 of its some 100,000 suppliers participated in the initiative, a far cry from the bold vision proposed back in 2008, when the goal was to track the environmental footprint of every product the firm sold.[75]

Nevertheless, the company continued to make big promises when it came to meeting environmental targets after launching the index. In 2017, for example, the firm announced Project Gigaton, setting an aspirational goal to eliminate a gigaton of greenhouse gases from its corporate system by 2030, and a year later, Walmart representatives traveled to the climate talks in Bonn, joining a host of American cities, corporations, and state governments in signing the "We Are Still In" declaration, pledging Walmart's commitment to the Paris Accord even as President Trump moved forward with plans to take the United States out of the agreement. Environmental groups and journalists alike gave the firm kudos for being a trailblazer in corporate greening initiatives.[76]

But independently verifying Walmart's progress toward ecological sustainability proved difficult because Walmart ultimately controlled the numbers. *Atlantic* reporter Orville Schell tried to dig deep in 2011, asking Walmart officials if he could have more details about factories that had been blacklisted from working with the discounter due to standards violations. "A clear answer was hard to come by," Schell said. And when he went further and asked whether Walmart auditors could offer greater detail regarding "high-risk" factories listed as being in the "Far East" on the company's audit sheet, he was simply told that the company did not "provide breakouts below the regional level."[77]

In 2018, Walmart expressed a similar ignorance in its audits for the Carbon Disclosure Project (CDP) run by a nonprofit that worked with corporations to collect and disclose greenhouse gas emission data. Walmart said only 670 of its more than 100,000 suppliers filled out a CDP survey, making it hard to calculate greenhouse gas emissions coming from its supply chain. Walmart noted that supplier "emissions were not verified by a third party" and that there was a high degree of variability in how different suppliers assessed their footprint. To fend off potential criticism, the firm offered an evasive excuse: "Calculating . . . emissions from purchased goods is complex and especially complex for the world's largest retailer."[78]

One of the biggest problems with Walmart's emissions reporting in the 2010s was that it did not provide a full accounting of greenhouse gases released by container ships carrying Walmart goods to market. Writing in 2013, author Stacy Mitchell explained the absurdity of this omission, noting that Walmart "now accounts for one of every twenty-five containers shipped to the U.S. and its imports are growing faster than the country's as a whole."[79] To truly measure Walmart's impact on climate change, oceanic transport had to be fully integrated into the calculus.

Here was a strange situation. A firm that had revolutionized open sharing of information between suppliers and retailers through satellite systems and advanced inventory management technology like Retail Link confessed it could not pin its suppliers down to get basic environmental data or offer full details about the ecological costs of getting its goods to market. Yes, the company had launched an ambitious sustainability index back in 2008, but a decade and a half later, only a fraction of their suppliers were included in that index and key data was still missing. The firm had the capacity to replace the smallest of trinkets on a store shelf in a matter of hours, activating a sophisticated digital communication network that stretched around the world, but when it came to environmental auditing, it seemed Walmart was still very much in the dark.

And it could afford to remain in the dark because no one else was watching. Just as there were no air monitoring devices in Williston to track emissions from idling cars waiting to turn left into the Walmart parking lot, so too were there no climate auditors keeping track of carbon emissions coming from ships transporting Walmart wares over oceanic waters. Once in motion, Walmart goods, and the consumers they drew to company stores, became difficult for regulators to track. Yet it was precisely the long distances both consumers and goods traveled that made the Walmart way of doing business so Earth-changing.

Despite all the fanfare about Walmart being a green business committed to curbing climate change, the truth is, the company has produced less than stellar results. According to its own self-reported data, Walmart's scope 1 (associated with Walmart stores and distribution center operations) and scope 2 (indirect emissions associated with energy purchases) greenhouse gas emissions increased by over 8 percent from 2006 to 2015. This does not even include scope 3 emissions, which cover Walmart's supply network, about which the firm still offers very limited data.[80] What would the numbers look like if Walmart included those scope 3 emissions?[81]

Sam Walton could have predicted this. What he learned in those early days of discount retailing was that firms would do his bidding only if they felt pressure to meet specific targets. He knew that he had to set the rules of the game, not the other way around. It was the "squeeze" that changed the retail world forever. But it worked only because he demanded that his suppliers open their books and show him how they made the sausage—or the shirt—so he could understand how far they could be pushed. With that knowledge he made businesses do the seemingly impossible. The motivation for him was economic, not ecological.

In the first decade of the 2000s, H. Lee Scott set the firm on a different course, designing an index that had the potential to put real pressure on suppliers to become more eco-friendly. A decade and a half later, though, the fervor that fired that mission had not produced rapid results, as less than 2 percent of Walmart's suppliers were participating fully in the index.[82]

This raises a question: Does the squeeze need to come from somewhere else in the years ahead? For far too long federal regulators have allowed Walmart to operate in the shadows of our economy because our environmental laws were never designed to regulate the firm's biggest impacts: those caused in its long-distance supply and distribution networks. The company has been largely free to set its own agenda for dealing with its growing greenhouse gas emissions and to set its own standards when it comes to meeting its climate change targets. It has largely avoided being squeezed to do the right thing. And the effects have been less than stunning, certainly not the kind of innovation that would have wowed someone like Sam Walton.

In many ways, it's not surprising that the EPA and other federal environmental agencies have done little to reimagine how stores like Walmart could be regulated in this new era of big-box retail. After all, since former Arkansas governor Bill Clinton came to the White House in 1993, the Democratic Party has rarely pursued radical expansion of environmental legislation. It might be argued that this is due to opposition from a Republican Party that since the 1990s has increasingly turned away from supporting environmental laws on the grounds that doing so would grow the size of government.[83] But Democrats bear blame as well, in part because the new brand of political liberalism Clinton and his predecessors espoused was based on neoliberal, free-trade policies that often gave businesses wide latitude to run their enterprises as they saw fit. The backlash against big government, a backlash spurred by whites' reaction against federal interventions in the Jim Crow American South during the civil rights movement, has meant that big boxes have been able to grow with few restrictions from the federal government.

But what would it look like if Walmart really felt the squeeze on sustainability, if watchdogs outside the company and in the federal government were empowered to force the firm to use its own sophisticated monitoring tools—tools developed through half a century of logistics innovation that enabled Walmart to ship goods to and from remote and rural corners of the globe—and commit to the promise the firm made back in 2008 to track the life cycle of all the goods they sell and to stop selling those items that are so harmful to the planet? That would indeed be a real retail revolution, one that could start in the logistics network an Ozark company built so many years ago that connected cities on one side of the world to rural communities on the other.

And that revolution would have to include by necessity another southern firm that by the 2010s had become a major partner with Walmart, helping the company ship a huge proportion of wares purchased on Walmart.com direct to consumers' doors.[84] This Walmart partner had adapted Delta's hub-and-spoke transportation model to transform express parcel delivery systems in America and hailed from the same delta region as the big airliner in a town that owed its first economic breaths to the Big Muddy: Memphis, Tennessee.

PART FOUR
The World on Time

Chapter Seven

I'M DOING THE SORT . . . IN THE SHADOW OF LENIN'S TOMB!

When turbulence rocked FedEx flight 88 late at night in December 1995, company systems engineer Chuck Noland was calm and collected. This was just part of the job. Shoes off and snoozing, he was jostled awake by the disturbance but quickly engaged in jovial banter with a copilot, Al Miller, who had been slumbering next to him, and then stumbled sleepily to the flight deck. They were over the Pacific Ocean, somewhere near the Cook Islands, headed to Asia.

The scene in the cockpit was a bit hectic. The captain, sweat beading on his head, was looking at a radar screen covered in green blobs that represented storm clouds immediately ahead. Lightning flashed, thunder rolled, and the plane rumbled as the flight crew anxiously tried to contact air traffic personnel thousands of feet below.

"Tahiti control, FedEx 88, broadcasting in the blind, how do you read? Tahiti control, FedEx 88, do you read?" Noland overheard the crew conversation and learned that they were now 200 miles off course, trying to avoid the tempest that swirled around them. He probably wondered whether they were going to make it on time to deliver company cargo. This was the stuff that mattered to Noland. He was a FedEx man, governed by the ticking of the clock. "The World on Time": so went FedEx's slogan, his mantra.

Despite the ruckus, Noland had faith in the folks flying the plane. He wandered back from the flight deck, grabbed his toiletry kit, and entered the lavatory, where he splashed some water on his face. Another rough flight, for sure, but he must have assumed all would end well. Nothing could prepare him for what happened next.

A piercing scream broke the silence of the lavatory chamber as Noland's body shot sideways, perpendicular to the floor, a rush of wind jettisoning him outward. Quick thinking was the only thing that saved him: he grabbed the frame of the bathroom door, holding on for dear life as he hung suspended in midair. The fuselage had exploded, and the cabin was quickly losing pressure from a gaping hole in the plane.

Oxygen. He needed oxygen. Fortunately, in that moment, Al Miller, was there, strapping a mask to his face so he could breathe. Everyone was panicked. The plane was falling from the sky and the pilots felt helpless. "We may have to ditch," said Miller, throwing an inflatable raft into Noland's lap. This was really happening. Though FedEx had never had a fatal crash in over two decades of flight, that statistic did not offer much assurance in this moment.[1] They were going down. Fast.

Seconds later, Noland realized the inevitable. Looking out through the windshield he saw a mountain of waves, an ocean in turmoil just feet below them. They were still flying so fast. The crash was going to be tremendous. "Brace for impact!" These were the last words he heard from the flight crew of FedEx flight 88.

So ends the opening montage in Tom Hanks's popular 2000 film *Cast Away*. In what some have dubbed free advertising for FedEx, *Cast Away* tells the fictional tale of company man Noland, who finds himself marooned on an uninhabited South Pacific island, the sole survivor of FedEx flight 88. When screenwriter William Broyles Jr. reached out to FedEx founder and CEO Fred Smith about doing the film, Broyles tried to be honest. "The good news is we want to use FedEx," he told Smith. "The bad news is one of your planes goes down."[2] After some debate, Smith and his marketing team decided to go along with the plan.[3]

And why not? In the end, *Cast Away* proved a remarkably effective tool for promoting the company brand. "As we stepped back and looked at it, we thought, 'It's not product placement, we're a character in this movie,'" explained FedEx's global brand manager Gail Christensen. This was better than spending thousands of dollars to have the company logo cleverly dropped in a few scenes. (FedEx did not pay Broyles or the production team anything for featuring the brand in the film.)[4]

Broyles wove a story that FedEx wanted told: a story of a scrappy southern firm from Memphis that had conquered the globe. As *Cast Away* went into production in the late 1990s, FedEx was excited about the prospects for growth in countries like China and Japan. The fact that Noland's plane is headed from Memphis to Asian markets matched well with reality. In 1995, FedEx opened a new superhub sorting facility in the Philippines, patterned on the hub-and-spoke delivery system Fred Smith had developed in Memphis—itself modeled on what Delta Air Lines had created for passenger flight in Atlanta.[5]

Other scenes in the film play up the firm's role as a disruptive force in international markets. Before Noland's fateful Pacific flight, viewers travel with him to Moscow, where FedEx seems to be on the front lines of replacing communism with capitalism. As a boy delivers a FedEx box, he walks past workers removing a placard of Vladimir Lenin. In case the message was too subtle, Noland helps unload packages off a FedEx truck in Red Square, the Kremlin towering in the near distance. He calls his girlfriend, Kelly Frears: "You're not going to believe this. I'm doing the sort . . . in the shadow of Lenin's tomb!" FedEx was creating conduits of capitalism that were changing the world.

Fred Smith was from a family that knew a thing or two about transport. His father, James Frederick Smith, was the proprietor of Memphis-based Smith Motor Company, a bus service he merged with Dixie Greyhound Lines in the 1930s. The company did well enough to afford Fred a comfortable life. Money from Dixie helped pay for Smith's preparatory school education as well as his tuition at Yale University. But though they were the foundation of his financial well-being, buses were not really Fred Smith's thing. He loved planes, a passion he developed in the Yale Flying Club in the 1960s.[6]

Aviation would play a big role in Smith's life when he finished college and headed off to Vietnam. The year was 1966 and President Johnson was ramping up the war. The young twenty-two-year-old started his first tour as a marine platoon leader stationed in Chu Lai, South Vietnam, but quickly rose through the ranks, ultimately becoming a forward air controller at Marble Mountain, just south of U.S. forces' coastal base in the oceanfront city of Da Nang. In this capacity, he coordinated the movement of troops, planes, and helicopters during air strikes. Naturally, timing mattered. Years later, Smith told veterans that logistics skills developed during these campaigns—skills that often meant the difference between life and death—became critical when he began to develop FedEx's business plan. Like many men from the American South, Smith was getting a real education halfway around the world in the 1960s.[7]

Two tours complete, Smith returned to the mid-South region and went into business with his stepfather, Fred Hook, who owned Arkansas Aviation Sales in Little Rock, Arkansas. Initially Smith serviced private jets, offering parts and repairs, but gradually he got into the business of taking used planes and reselling them to wealthy patrons.[8] By the end of the 1960s, ready for something more, he reached out to a small group of venture capitalists with an idea he had been mulling over for some time.[9]

The beginnings of an idea for something like FedEx had first come to him while he was still a student at Yale. There, just twenty-one years old, he penned a thesis showing that there was a market for small-package express air delivery, something that, in Smith's recollection, did not wow his professors. "I don't know, probably made my usual C," he told a reporter years later.[10] Nevertheless, the idea marinated through his years in Vietnam and Arkansas, and in 1971 he decided to turn it into a reality.

Unlike Coca-Cola founder John Pemberton, who was cash-strapped and struggling in the Reconstruction-era South, Fred Smith was a wealthy young man in the booming Sunbelt, the heir to a Dixie dynasty that left Smith and his two sisters access to roughly $14 million. Smith tapped these funds to incorporate his air delivery business on June 18, 1971. He called the new firm Federal Express because he intended to use his jets—just a couple of Dassault Falcons at first—to transport checks and documents between various banks and the Federal Reserve's branch offices.

However, the Federal Reserve ultimately balked at Smith's proposal, forcing Smith to turn to the private sector to drum up business.[11]

This was no easy task. At the time there were other firms, most notably the Railway Express Agency (REA) and Emery Airfreight, that had long worked with the major airlines to ship packages and freight. REA and Emery were known as forwarders because they did not actually own and operate their own fleet; instead they offered logistics services and ground support to airlines looking to fill cargo space on their planes. If Federal Express was going to make money, it was going to have to beat out REA and Emery, firms Smith came to call the "enemy" in company meetings.[12]

Fortunately, Smith had a major advantage. Because REA and Emery depended on passenger airlines, they could not ship cargo whenever they desired. Most passenger flights took place during the day and early evening, meaning overnight shipments of packages would be virtually impossible. But shipping overnight was the key to a successful express air freight service. Smith knew this because in 1972 he commissioned A. T. Kearney, a consulting firm based in Chicago, to investigate whether there was a market for Federal Express. The Kearney analysis revealed that most businesses interested in air transport wanted to get time-sensitive documents and goods out the door late in a business day and have them arrive early the next morning. REA and Emery simply could not offer this arrangement to its clients given the strictures of passenger flight schedules.[13]

But even if Federal Express had an opening, it was going to need help if it was going to get off the ground, and it got that help from the federal government. Still operating out of Little Rock, Arkansas, Federal Express secured a contract from the U.S. Post Office to offer mail service to small rural communities in July 1972. These mail contracts, the first real moneymakers for the company, brought in an estimated $300,000 monthly, a big deal for a fledgling firm. As late as summer 1973, the firm was making an estimated 90 percent of its revenue from mail deliveries.[14] At the same time, Smith found another way to turn federal dollars into private capital, setting up a flying school for Vietnam veterans that was designed to get them certified to captain company planes. The GI Bill covered costs, meaning the Veterans Administration was essentially financing early Federal Express workforce training. As with many other businesses that flourished in the postwar American South, government spending proved crucial to the birth of Smith's airborne company.[15]

By the end of 1972, still searching for big corporate contracts but nevertheless afloat thanks to airmail deals, Smith was committed to expanding his business. In September of that year, he had secured a key exemption from Civil Aeronautics Board (CAB) regulations that allowed him to fly wherever he wanted with company freight, so long as the total weight of his loaded-down planes remained less than 7,500 pounds—1,000 pounds less than the maximum carrying capacity of company Dassault Falcons. This would allow Federal Express to fly air routes once reserved for just a few American airlines. Smith's southern firm greatly benefited from new CAB rules in the 1970s that gave Federal Express access to a world well beyond the South.[16]

In this moment, Fred Smith began to lay out the plans for a delivery system that would forever change the global economy. Smith was trying to do for packages what Delta had done for passengers by adopting a hub-and-spoke transportation system. Now that he had federal approval to ship documents and goods far and wide, Smith wanted to initiate construction of a grand sorting facility. The idea was that planes from cities across the country would all arrive between 10:30 P.M. and 1 A.M. at this central hub, where packages would be unloaded from incoming jets, sorted, then reloaded on outbound jets heading to destination cities early in the morning. In and out, all overnight—that was the vision Smith had for his delivery business.[17]

But where should FedEx locate this proposed node of the hub-and-spoke system? Fred Smith came to see that the airport in Little Rock, Arkansas, could not support his company if his firm got bigger. He also did not like the fact that the Little Rock airport was often shrouded in fog and subject to bad weather. It just did not have the attributes he needed. As a result, he gathered a few company men, including consultants he had worked with at Kearney, and they surveyed the country in search of the perfect aviation hub. He found it in the town that had made his dad's firm famous so many years ago: Memphis, Tennessee.[18]

Environmental factors played heavily in Fred Smith's decision to locate Federal Express's first superhub in Memphis. Smith, who had grown up in Memphis, was well aware of the advantages this city offered his firm. He knew that to stand out Federal Express had to deliver packages on time every time, but to do that, the firm had to avoid delays caused by bad weather. Here, Memphis offered many benefits. Climate was key. Far enough south to avoid icy conditions and blizzards that routinely plagued cities farther north, Memphis was also several hundred miles distant from the coastal storms and hurricanes that caused nightmares for Gulf Coast cities. In addition, Memphis was far enough west to permit timely direct flights to the Pacific Coast metropolises of Los Angeles and San Francisco but still close enough to the Atlantic seaboard to allow swift transfer of goods from New York, Atlanta, and Miami.[19]

But big cities were not the real draw for Federal Express. What made Memphis special was that it was, as FedEx general manager Roger Frock explained, "at the center of gravity of the small package market."[20] What Frock meant was that FedEx's small planes in Memphis were able to reach American towns that many passenger airlines were ignoring. "Eighty percent of the small priority shipments originated in or were destined for cities outside the country's twenty-five largest markets," Frock noted, citing rigorous research completed in the early 1970s. At that time "airlines were decreasing service to the smaller passenger markets," Frock said, creating a huge opportunity for FedEx.[21]

Access to muddy waters as well as clear jet streams made Fred Smith's dream possible in Memphis. After all, another critical natural asset that made this Tennessee town attractive when it came to trade was the Mississippi River. In the Gilded Age, Memphis had become a central trade hub, and by the end of the twentieth century it remained the nation's largest inland river port on the Big Muddy. City founders in 1819 knew the Mississippi River made this city, naming their town after the once-great ancient Egyptian capital located on the Nile River just south of Cairo. The message was clear: this was to be an empire built on water.[22]

The river was critical to town commerce. Cotton floated through Memphis, making the city home to the largest spot market for cotton trading in the world by the turn of the twentieth century. In the Gilded Age, as railroads connected Memphis to far-distant southern forests, a robust lumber trade developed in the area. City boosters billed their town, with its many lumber mills humming along the riverfront, the "Hardwood Capital of the World." Memphis was an organic city, transforming agricultural staple crops and southern timber into commodities that drew financial capital. With this money, in 1929 city developers completed construction of the town's first municipal airport, the critical piece of infrastructure from which Fred Smith would launch his aviation business more than forty years later.[23]

In the 1970s, the Memphis airport offered certain advantages beyond just favorable climatic conditions. Sited on flat farmland south of the city, with the Mississippi flowing to the southwest, there were no major waterways or hills hemming in runways. Looking out over the flat horizon surrounding the airport, Fred Smith must have imagined ample prospects for growth. Tarmacs ran north–south as well as east–west and had sufficient instrumentation to support all-weather takeoffs in all cardinal directions, something the Little Rock airport in Arkansas (another FedEx-headquarters candidate) did not.[24] For a company that cared about minutes and seconds, the multidirectional layout of the airport offered major advantages for timely arrivals and departures.

The political and economic environment in Memphis also appealed to Smith. One of the biggest strengths was the nonunion, low-wage workforce in the area. Fred Smith hated unions and fought any organizing efforts initiated by employees. By the 1980s, FedEx would be the largest employer in the state of Tennessee, so keeping labor costs down was a major priority among C-suite executives.[25] Smith's business model hinged on savings accrued by hiring part-time contract workers, and Memphis had an ample supply of such labor. In 1979, the entire 5,300-person FedEx team—pilots and packers—remained unaffiliated with labor unions.[26]

The antiunionism found in Memphis had a long history. As we have seen, in the 1940s and 1950s many Black and white farmworkers left a southern countryside increasingly controlled by New Deal–funded landholding elites who were rapidly mechanizing their farms. Many of these southern migrants headed to cities like Memphis, where they experienced exploitation at the hands of employers now flush with workers. But that exploitation often failed to spur strong union organizing, in large part because of the Jim Crow racial caste system, which ensured that efforts to bring about interracial solidarity around labor issues never materialized. To be sure, during the civil rights movement, Memphis citizens fought back against labor exploitation, perhaps most notably in the sanitation worker strikes of the mid to late 1960s, which culminated in 1968 when Martin Luther King Jr. was assassinated at the Lorraine Motel in Memphis. And some industrial firms that had moved their factories to Memphis in the 1950s and 1960s began to question whether the cheap labor market that had enticed them to the state of Tennessee was destined to remain in the years ahead. But even after the civil rights movement, Memphis continued to be, as historian Wanda Rushing put it, a city in which "racial identity has proved to be a barrier to class identity and to labor organizing, hence, an economic benefit to employers."[27]

Memphis also had a city government eager to provide financial incentives to attract business. The municipality issued $3 million in tax-exempt bonds in 1976 to help finance the construction of a new terminal for FedEx.[28] And beginning in the 1980s, the city's Industrial Development Board offered companies like FedEx the opportunity to participate in a Payment in Lieu of Taxes (PILOTS) program, which allowed FedEx to pay a fee rather than its full tax liability on assessed property. Such tax assistance proved extremely lucrative for firms like FedEx.[29] As historian Jim Cobb has shown, this was standard fare among cities that were actively "selling the South" at this time. Weak labor laws and lucrative financial incentives were key parts of the financial formula designed to lure businesses to southern cities, and Memphis played these cards well.[30]

By FedEx's Launch Day, March 12, 1973, the company thought it had found the ideal habitat to nurture its corporate organism. Smith now watched anxiously as he pumped the heart of his commercial system and watched its arteries flow for the first time.[31]

It was an absolute disaster. The company had anticipated shipping hundreds of parcels, but in the end, the grand total for the night was astonishingly low: just six packages shipped. The customers FedEx had been courting simply did not come through with orders. The system was just too new. Everyone felt defeated. For months prior to Launch Day, Federal Express salespeople had spread out to drum up business for the company. These salespeople focused on ten cities within striking distance of the company hub, including seven southern cities: Atlanta, Cincinnati, Dallas, Greensboro, Jacksonville, Kansas City, Little Rock, Memphis, Nashville, and St. Louis. Reports from the sales force suggested the company would have ample business, but as planes came back to the Memphis hub on March 12, it was clear things had gone horribly wrong.[32]

FedEx's financial position was dire. According to company COO Roger Frock, the company was down to its last $5,000 by July 1973, and only survived due to a lucky hand Fred Smith had at a blackjack table in a Vegas casino that turned $5,000 into about $32,000. For the moment, Smith had gambled and thereby escaped disaster, but trouble loomed ahead.[33]

In terms of natural resource availability, it was a terrible time to face such a business blunder. Just a few months after the Memphis move, the OPEC oil embargo began, creating fuel shortages that rocked FedEx as much as it had Delta. FedEx was particularly hard hit in this time of financial crisis, not only because it meant fuel costs were on the rise but also because Smith was busy trying to drum up capital from investors who did not necessarily look favorably on investments in a fossil fuel–driven freight service.[34]

In this moment, Fred Smith went to Washington to find relief, and as he would do so many other times in his career, he leveraged connections to southern lawmakers to get special deals that helped his firm survive. Following the OPEC embargo, the Nixon White House imposed fuel quotas on American airlines, essentially holding them to 1972 consumption levels. Because Federal Express was just a small Little Rock startup in 1972, this restriction would have buried any chance the firm had of making a recovery. Facing this existential threat, Smith secured a meeting with Tennessee-native Howard Baker Jr., then early in his senatorial career and over a decade away from serving as Ronald Reagan's chief of staff. Baker helped Smith get a conference with the Nixon appointee in charge of oil quotas, wherein Smith played up his Vietnam service with the former military man, gaining exemptions that enabled Smith to get more fuel for his fleet.[35]

More good news was on the way, this time in the form of bad news for one of the firm's rivals: the United Parcel Service (UPS). UPS, founded in Seattle in 1907, was a lot older than Federal Express and had an incredible ground game, complete with a massive trucking network that dwarfed Smith's firm. But in the 1970s, UPS had not really invested in a fleet of planes that could serve the overnight delivery market. It coordinated shipments of parcels on commercial airlines, but it was way behind FedEx in its air game. UPS would not start express overnight service until 1982, when it finally recognized that it had to focus more on jet streams if it wanted to prevent FedEx from eating into its revenues.[36] Nevertheless, in the 1970s UPS's two-day express service still offered competition to Fred Smith's company, meaning what was bad for UPS was good for Federal Express. So when the Teamsters Union initiated a strike in 1974 that stalled UPS shipments around the country, FedEx went on the offensive. Unlike UPS, which had a large union workforce, FedEx did not have a single union employee, meaning the strike did not slow it down in any way. The firm used this advantage—an advantage in many ways born of its southern origins—approaching courier customers once serviced by UPS.[37]

After the initial launch failure, Federal Express had slowly increased package deliveries, and the firm gained a reputation for prompt delivery and quality service. A. T. Kearney's early analysis was right: there was a real market for overnight express air delivery service, and FedEx had tapped it. By the summer of 1975, for the first time in Federal Express's short history, the company was finally in the black. A year later, Smith could boast that the firm was shipping 16,000 packages a day, reaching approximately 10,000 cities and towns across the country. Company president Arthur C. Bass told reporters, "We don't see anything in the foreseeable future that will put us back into the red."[38]

But despite the optimism, there was a real problem that stood in the way of future growth. Under CAB regulations, FedEx was still restricted to flying small planes with a mere 7,500 pounds of capacity; any bigger and the firm would have to seek certification for routes and abide by shipping rates set by the government. Bass told the press that the company was having to make multiple flights to small cities just to cover the company's current package volume. If it had to continue to rely on its fleet of Dassault Falcons, now numbering just over thirty, it was going to expend a lot of extra fuel, which would mean a lot of lost revenue. But all this became a nonissue when, late in 1977, after much FedEx lobbying, Congress passed the Air Cargo Deregulation Act, "virtually eliminating," as FedEx explained in its 1978 annual report, "all government economic regulation of the movement of goods by air in the continental United States." This law targeted the air cargo industry while the Airline Deregulation Act of 1978 focused on passenger airlines. The company immediately moved to get bigger planes, purchasing ten Boeing 727s in 1978. That same year, hoping to capitalize on the recent good news, Fred Smith decided to take his company public. The initial public offering in April was a tremendous success: the company's stock price had nearly doubled by August, and the *Washington Post* declared FedEx the year's "hottest stock issue."[39]

In the 1980s, FedEx initiated major expansions into international markets. The company purchased companies in Europe and specifically eyed Asia as a particularly enticing area for growth, but the issue was that the company did not have aviation rights to fly its own planes into Japan and China. In 1989, Fred Smith acquired Los Angeles–based Flying Tiger Line, one of the few American air cargo shippers that had been certified for travel to airports in Japan, Hong Kong, and other key ports of entry on the other side of the world.[40] Founded in the mid-1940s by World War II veterans who had served in the Flying Tigers squadron, which had fought in China, this air cargo company had long flown flights across the Pacific to Asian markets. The purchase was a big move for Smith, one that came with an $880 million price tag.[41]

Wall Street investors did not necessarily welcome the Tiger takeover. Just months before the acquisition was finalized, one financial reporter wrote, "Federal Express Corp.'s efforts to gobble up business overseas have left the air express concern with a bad case of indigestion."[42] He pointed to FedEx's struggling financial performance in the second half of the 1980s. In just the first half of 1989, for example, losses from international operations totaled more than $30 million. Would the Tiger deal cause more headaches for FedEx?[43]

Several factors played into FedEx's less-than-stellar performance during Ronald Reagan's second term in office. Perhaps most important was that Europe was a disaster for the firm. Company executives simply could not turn the multitude of acquisitions the firm had made in Europe into a sleek, unified delivery system. By 1992, the company had decided to halt its "intra-European service," telling its shareholders that it had "discontinued these operations" because the "intra-Europe and intra-country market potential simply did not justify our remaining in these segments."[44]

There were other problems, including ZapMail, an "electronic document transmission service" introduced in 1984, which was pretty much a fancy fax machine system. The company invested lots of capital in its ZapMail system only to find that businesses were more than happy to use cheaper alternatives. Though FedEx had tried to head off problems associated with a disruptive technology, the reality was that electronic document transmission systems represented an existential threat to FedEx. "Its very mode of delivery is being challenged by new technology, most notably facsimile machines," the *Wall Street Journal* explained in 1988.[45] Here was a new market predator that could devour FedEx's business. The company had to adapt to a new commercial ecology or die.

FedEx's response to the fax threat had implications for the environment. To stay in business, the company was going to have to push more than paper; it was going to have to ship stuff: computer parts, perishable goods, and so on. To be sure, the company had been shipping packages as well as documents from its founding, but in an era where time-sensitive papers and business memos could be transmitted in minutes via phone lines, FedEx was going to have to replace lost business in this sector with new parcel types. Fortunately, the company, freed from the CAB restrictions of the pre-1970s, now had big planes that could carry bigger cargo. Of the more than 380 aircraft in its fleet in 1990, 89 were sizable Boeing 727-100s and 19 were Boeing 747 "Jumbo Jets."[46] FedEx had the capacity to carve much wider commercial arteries than it once could with its original thirty-odd Dassault Falcon private jets. Where it went, it could accelerate the transformation of natural capital into consumer goods.

In Asia, FedEx turned a major problem into an asset. Rather than be undone by electronic gadgetry coming out of Japan, Taiwan, and other Asian nations, FedEx would become the transporter of the circuit boards and semiconductors that made fax machines and computer transmitters work. The digital age was coming; in fact, it had already arrived. Embedding itself in the commodity networks that connected this new interconnected world was a smart move for this firm that had feared fax machines.

But to become the commercial player it wanted to be in Asia, FedEx had to find its Memphis there: a superhub that could serve as the ideal sorting facility for what FedEx hoped would be an enormous amount of trade. As Fred Smith and company leadership pondered its options in the early 1990s, a natural disaster in Southeast Asia was about to reshape the future of the business.

Chapter Eight

WHAT WE WANT TO DO IS CUT THE SUPPLY CHAIN

On the afternoon of July 15, 1991, Mt. Pinatubo in the Philippines erupted. For the few U.S. Geological Survey (USGS) scientists and military personnel left behind at Clark Airbase less than nine miles distant, the scene that lay off on the horizon was nothing short of apocalyptic. Volcanic ash shot up more than twenty miles as a dense gray cloud began to billow across the land, smothering hills and valleys, burying some areas below more than 600 feet of spewed debris.[1]

The second-largest volcanic eruption to occur in the last century, the Pinatubo blast was so big that it changed Earth's climate. Satellite images showed ash particles traveling around the planet several times, pushed along by the same jet streams that helped FedEx planes speed through their daily routes. An estimated 20 million tons of sulfur dioxide entered the atmosphere during the explosion, blocking out enough sunlight to reduce average global temperatures 1 degree Fahrenheit between 1991 and 1993. This was the reverse of global warming, a fact noted by geoengineers who hence proposed controlled releases of sulfur dioxide from balloons suspended from hoses as a solution to the climate change crisis. The "Pinatubo Option," they called it.[2]

The blast was only part of the bad news. In a remarkably unfortunate set of circumstances, Typhoon Yunya came barreling across the South China Sea at precisely the same time as volcanic ash spewed upward. Typhoon rains turned the gray matter in the sky into dense clumps of heavy mud that crushed buildings and homes. The torrent also triggered mudslides that inundated towns and villages.[3]

In a matter of hours, Clark Airbase lay in ruins and the U.S. Naval Air Station at Cubi Point in the deepwater port of Subic Bay some twenty miles south of Pinatubo lay buried beneath roughly a foot of volcanic ash. Coupled with Clark Airbase, the navy's Subic Bay base was the United States's largest overseas naval outpost in the world and had been a central staging ground for American troops seeking to curb the "Red Tide" of communism in Southeast Asia during the Cold War. A young Fred Smith visited Subic Bay en route to fight the Vietcong in Vietnam. Now, some twenty years later, Smith would not have recognized the place. This was a doomsday landscape.[4]

The U.S. military hoped to turn disaster into opportunity. At the very moment volcanic ash fell on Cubi Point, negotiations between Philippines government officials and U.S. diplomats over renewal of a decades-old treaty that gave U.S. forces dominion over military assets in Subic Bay became deadlocked. The U.S. lowballed the Philippines government when they made their renewal offer in 1991, citing the recent damage to the facility caused by Pinatubo and Yunya.[5] Filipino negotiators, already emboldened by impassioned patriotic speeches made by charismatic politicians in the Philippine Senate, did not accept the offer. Calls came to end American colonial occupation, and in September 1991, the Philippine Senate voted to reject treaty renewal.[6] Thousands of American GIs and their families packed their bags. Roughly fifty years after a routed General Douglas MacArthur told the Filipino people, "I shall return," most American military men in Subic Bay were heading home.[7]

In this moment of natural disaster and resurgent Filipino nationalism, FedEx looked to strike a deal. Remarkably, within a matter of months, much of the volcanic debris that had been dumped on Cubi Point had been cleaned up, and FedEx began to eye the facility as an ideal location for a new superhub that could give the firm easy access to Asian markets. Here was the Memphis of Southeast Asia. Politically, the timing was good. Filipino president Fidel Ramos had just been elected in 1992 on a platform that promoted free trade and public policies designed to entice multinational investment in the Philippines.[8] In the fall of 1993, while visiting U.S. officials in Washington, D.C., Ramos met with FedEx officials who made a pitch to take over the former naval base in Subic Bay. A year later, FedEx inked an agreement that gave the firm control over the naval installation. Ramos crowed that the Philippines was "back in business at the heart of Asia."[9]

Part of what made Subic Bay so attractive was its political and economic climate—a commercial environment that in many ways matched what Smith liked about Memphis. FedEx's new hub was in the middle of a pro-business enclave designed by Filipino politicians to attract multinational investment. Dubbed the Subic Bay Special Economic and Freeport Zone, this was a duty-free commercial area with exceptionally low income tax rates and lax customs enforcement. Under the Bases Conversion and Development Act of 1992, the Subic Bay Metropolitan Authority (SBMA) was given broad authority to negotiate deals with investors and make decisions about local governance. "Exemption from all local and national taxes with only a 5% corporate tax on gross income," one investor-focused SBMA brochure read.[10]

Richard Gordon, mayor of Olongapo City and head of the Freeport Zone, promised businesses that there would be "good industrial co-operation between labor and management." Gordon was a larger-than-life political figure in the area, credited with forming an army of volunteers who looked after the Subic Bay property while its future was in limbo. For outsiders looking in, Gordon was a man with considerable power, "a godfather-like figure" as one person put it, so it was reassuring when he told investors that there would be "good industrial peace" in this place.[11]

In short, here was a pro-business political environment within easy reach of both China and Japan that was geographically situated at the heart of Asian markets.

In 1995, FedEx officially started operations at its AsiaOne hub in Subic Bay. That year, FedEx acquired Evergreen International Airlines, the only American air cargo company with rights to ship to China. FedEx now had sole access to an air express market in the most populous country in the world.[12] The Subic investment now looked even better.

Except for one big problem: Japan. Just as FedEx gained access to China, Japan threatened to restrict the company's access to its markets, arguing that it wanted to renegotiate aviation deals that diplomats found discriminatory to Japanese carriers. Smith was upset. At the same time that his multimillion-dollar Subic Bay project was coming online, a critical link in FedEx's international network was in jeopardy. He had to do something to make sure he did not lose Japanese business.[13]

The fact that Fred Smith and FedEx hailed from the American South mattered when it came to breaking the Japanese stalemate. Former Arkansas governor Bill Clinton and Tennessee politician Al Gore now occupied the White House, and they both had homeland roots that connected to FedEx. Smith's aviation investments started in Arkansas and matured in Tennessee. Naturally, as the founder of one of the booming businesses in Gore's home state, Smith had come to know the vice president when Gore was still a young senator on Capitol Hill.[14] Clinton, for his part, had always been concerned about southern firms from the Mississippi Delta. At "Delta Visions, Delta Voices: The Mississippi Delta beyond 2000," a national conference organized by the White House in 2000, President Clinton offered opening remarks, saying, "I'm here today to pledge that the federal government will be in the forefront of . . . working with the people of the Delta; and that I personally will be working" to help communities in the Delta "until the day I leave office and for the rest of my life." Fred Smith, one of the delta's most dynamic businessmen, was in attendance, an invited guest who spoke on a panel later in the day.[15] He was, as Clinton acknowledged in another speech that year, someone the president had known "for many years."[16]

In February 1996, President Clinton received a written request from the chair of the Democratic National Committee to reach out directly to Fred Smith for a party donation. "Ask him for $150,000," the memorandum said. White House phone records subsequently revealed that the president did try to call Smith. Whether he made the pitch or not is unclear, but campaign finance records show that FedEx gave $435,000 to Democratic senatorial candidates and the DNC in just three months of the summer and fall of 1996.[17]

At the time, Fred Smith had good reason to want to curry favor with Clinton. The Japan situation was not getting any better, and FedEx executives were getting frustrated. They had got their new Subic Bay sorting hub up and running, but without open access to Japan airstrips, FedEx's expansion into Asian markets would be labored at best. The firm was losing tens of millions of dollars because of this.[18]

Perforce, Mr. Smith went to Washington, securing a private lunch meeting with the president on August 23, 1996. This was very unusual. White House special counsel Lanny J. Davis admitted that Thomas F. McLarty, the president's former chief of staff who organized Smith's visit, "rarely set up meetings between the president and business leaders." The optics of such private meetings with company CEOs were bad, something the president generally avoided in order not to appear captured by big business. These were tense times, and such a move could jeopardize political futures. Clinton was midway into a nail-biting election season following the Republican congressional surge of 1994. He had every reason to avoid controversy; nevertheless, he sat down with Smith for roughly forty-five minutes to hear what was on his mind.[19]

Smith made his pitch. He wanted Clinton to impose stiff sanctions on the Japanese to force compliance with provisions of a 1952 U.S.-Japan aviation treaty that would have allowed FedEx to expand aggressively in Japan. Clinton listened, and expressed his sympathies for FedEx's plight, but according to Smith, he simply would not move forward with immediate sanctions. Doing so would be too risky. Clinton promised to return to the matter after he had defeated Dole in November, but it would have to wait until after the election. Following the meeting, Smith told *Washington Post* reporter Bob Woodward that Clinton was being "lily-livered."[20]

But even if Smith was not satisfied, Clinton officials fought hard for FedEx in its negotiations with Japan. Administration negotiators simply would not budge on any concessions to the Japanese unless they agreed to give FedEx the open access to airfields they desired.[21]

Almost a year after his secret meeting with Clinton, Mr. Smith went back to Washington, this time testifying before Congress on the Japan issue, which remained unresolved. He was exasperated. "I have seen everyone," he grumbled to Congress. "I have seen the President, I've seen Secretary Christopher. I've seen Under Secretary Spiro, Secretary Pena . . . I could list them on and on." Yet, "here we are four years after the

Japanese first declared aero-political war on FedEx without resolution." It was time to act.[22]

Smith had the ear of those in attendance. For years he and his firm had spent millions on the campaigns of those gathered. In the lead-up to the congressional elections of 1994, FedEx doled out roughly $850,000 in political campaign donations to over 200 politicians on Capitol Hill, both Democrats and Republicans. In 1995, the firm ranked third among all corporate donors to political action committees, spurring the *New York Times* and the *Wall Street Journal* to run exposé stories detailing how FedEx had become a "major force" in Washington.[23] Beyond money, FedEx also used company aircraft to fly politicians hither and yon. Nobody wanted to upset a firm that helped them connect to constituents across the country—no one except perhaps Senator Feingold, who explicitly called FedEx out in his passionate arguments for campaign finance reform.[24]

After Smith's speech, Clinton trade negotiators kept at it, situating FedEx's concerns front and center in negotiations. It took until February 1998, but Clinton administrators finally struck a deal, one that secured the air rights FedEx had wanted. The company was now ready to conquer new markets on the other side of the world. And so too was Delta Air Lines, which had gained access to more routes to Japan because of the new aviation pact. In short, two southern companies eyed major expansion in Asia thanks to help from a southern president who had kept Federal Express's wishes in mind when negotiating a major world trade deal.[25]

Asia proved an important new frontier for FedEx. In an era when email was eating away at the company's document delivery business, FedEx embraced electronics, forming key partnerships with semiconductor manufacturers and computer businesses with manufacturing assets in Asia. Acer, a Taiwanese-based computer supplier, became a big client, moving a key manufacturing facility to Subic Bay so that it could be closer to FedEx's hub. Acer was one of the largest personal computer manufacturers in the world, and the company relied on FedEx's swift shipping system to stay relevant in a technology industry where the pace of innovation and change was frenetic.[26]

National Semiconductor, a U.S. firm with manufacturing plants in Singapore, also became a big client.[27] An e-commerce briefing paper marked "not for distribution" delivered to President Bill Clinton in 1998 noted that FedEx had helped National Semiconductor achieve a "reduction of its average customer delivery cycle from four weeks to one week."[28] Philips semiconductor company, one of the largest chip makers in the country and a key logistics client of FedEx since 1997, relied on FedEx's Subic Bay hub to get its semiconductors from Bangkok production facilities to markets in Europe and the United States.[29] FedEx was accelerating the pace of capital flows from Asian manufacturers to U.S. markets.

In so doing, FedEx was hastening ecosystem transformation in Asian communities. Scientific studies in the 1990s and 2000s showed that semiconductor manufacture yielded a host of toxic pollutants—including arsenic, benzene, lead, and cadmium—that could affect air and water quality.[30] According to *Quartz*, Santa Clara County, California, the birthplace of the semiconductor industry, became "home to more toxic Superfund sites than anywhere else in the country" because of decades of semiconductor waste generation in the area.[31] But by the 1990s, a lot of this pollution had been dumped on Asian manufacturing partners that now could supply American clients via express air freight services provided by companies like FedEx.[32]

The people most affected by the toxic compounds associated with semiconductor manufacture in Asia were workers, mostly women, laboring in overseas factories in countries with lax environmental and labor regulations. Back in the 1980s, epidemiologists at the University of Massachusetts Amherst and the University of California at Davis discovered that the chemicals used in the semiconductor industry wreaked havoc on women's reproductive systems. After years of research, scientists showed conclusively that women working in wafer production and packaging were having miscarriages at a rate twice the national average. Industry was stunned, but instead of fixing the problem, they sent it overseas. Citing "confidential data" leaked to its journalists in 2017, *Bloomberg Businessweek* reported that "thousands of women and their unborn children [in Asian production centers] continued to face potential exposure to the same toxins [that had affected Silicon Valley women workers years earlier]." These were high costs in a high-tech economy.[33]

And semiconductor plants also had major effects on hydrological systems. According to some estimates, factories used over 2,000 gallons of water to produce one thirty-centimeter semiconductor wafer. As a result, big production facilities in China and Malaysia used millions of gallons of water every day to produce their chips. This resource demand might have proved a major limit to growth had the center of semiconductor production remained in water-stressed California, but thanks to the new conduits of global capitalism created by FedEx and other express air cargo companies, firms could now get the water they needed for computer chips in Asia, which by the 2010s was home to over 75 percent of the semiconductor industry.[34]

FedEx was not directly involved in running semiconductor plants, but by giving these plants access to its express delivery network, it helped them expand their production capacity and thus their ecological footprint. And electronics was just one area where express air freight service was changing the geography of global trade. In 1996, FedEx coordinated over sixty-seven nightly landings on the Subic tarmac at its AsiaOne hub, "ferrying everything from Timex watches to fresh tuna around Asia," according to the *New York Times*. A year later, the company was shipping approximately 2.8 million packages each day from its new hub.[35]

As the *Times* piece revealed, FedEx's ecological impact stretched beyond land to oceans, as frozen fish and other marine life went flying on FedEx flights to consumer markets around the world. To be sure, FedEx was not the main player in this fish industry; Japan Airlines had filled its cargo hulls with fresh tuna going back to the 1970s, and other passenger carriers held considerable sway in the express air delivery of fish. But FedEx nevertheless played a role in making sushi available to consumers in middle America and beyond, which in time placed increased stress on already depleted fish populations in the Pacific, especially prized bluefin tuna schools.[36]

Ocean waters proved both an asset and a hindrance to FedEx in Asia. By the mid-2000s, FedEx began to face natural barriers to growth as the company strained to service expanded Asian trade from its Subic Bay hub. A major problem was the airstrip. FedEx could not fill its biggest planes to maximum capacity for transpacific flights because doing so would make the planes too heavy for safe takeoff on the short Subic Bay runway. Problems also occurred during landings. At one point a FedEx plane, fully loaded with cargo, slid off the runway and into Subic Bay, unable to decelerate before reaching the end of the airstrip.[37] This was a costly problem, but to solve it by expanding the tarmac was cost prohibitive. The airstrip was hemmed in by hills and water. There was simply nowhere to go except out to sea, and doing so would be expensive. To make matters worse, Filipino president Joseph Estrada, elected in 1998, began making aviation deals that favored Philippines Airlines over foreign competitors.[38]

It seemed the political climate was turning toward protectionist policies that Fred Smith had always hated. It was time to get out. In 2006, FedEx announced it would be moving its AsiaOne hub to Guangzhou, China. In 2009, the firm made the move.[39]

In the years ahead, FedEx eyed a booming pharmaceutical market in Asia as a key area of potential growth for the firm. FedEx specialized in delivery of high-value, time-sensitive products, precisely what pharmaceutical and biotechnology businesses produced. The company could offer drug firms attractive services, such as temperature-controlled delivery systems that ensured that vaccines and medicines did not denature en route to customers. By the mid-2010s, the company began investing heavily in "cold chain" technology that could keep pharmaceutical products at low temperatures along their entire route of travel. In an era of climate change, FedEx was creating a cold commercial environment that could insulate sensitive commodities from the heat of a warming world.[40]

These investments saved lives. FedEx served customers such as Japanese-based Chugai Pharmaceutical, a company that produced, among other things, the widely used flu vaccine Tamiflu, helping to connect this business to communities trying to control contagion.[41] During periods of crisis, the firm also used its transportation network to do good around the world. In 2005, after tsunamis made landfall in Sri Lanka and Indonesia and Thailand, FedEx flew relief aid to people on the ground. Likewise, when earthquakes ravaged Myanmar in 2011, FedEx rushed much-needed medical supplies to victims there.[42]

But if FedEx channeled lifesaving goods to Asian communities in the wake of natural disasters and connected communities to pathogen-fighting vaccines, not all drug shipments FedEx made were benign. In 2018, U.S. senators testified that they had evidence that opioid manufacturers used FedEx to get drugs like fentanyl into the country. Such drugs were primarily manufactured in China and would have, in all likelihood, come through the company's AsiaOne hub. FedEx was one channel among many that distributors used to get opioids into the United States, with most drug sellers preferring the U.S. Postal Service because they assumed packages would be less scrutinized. But private express shippers like FedEx were caught up in the opioid crisis, and the firm's executives made clear that they wanted to cooperate with authorities to identify suspicious packages and do their part to help stop a public health crisis.[43]

FedEx's involvement in the debate about opioids revealed how far this Memphis company had come since its humble beginning in the 1970s. What federal regulators recognized was that FedEx had literally become a gatekeeper of the global economy. FedEx could play a big role in improving public health by making tough choices about what firms it did business with and what goods it shipped around the world.

This was something environmental activists recognized in the mid-2010s when they called for a FedEx boycott because of the firm's alleged involvement with illegal shipping of shark fins. Long considered a delicacy reserved for soups served at special occasions in Asian markets, shark fins became the target of conservation groups in the 2010s that worked to expose the gruesome fishing practices that brought fins to market. A bloody affair, the supply-side story of this delicacy was disturbing: fishermen were dumping sharks severed from their fins back into ocean waters, where they writhed and flailed until they bled out and died. In 2016, WildAid, a California-based conservation group, estimated that over 70 million sharks were killed each year to feed consumer demand for shark fins.[44]

Hong Kong has long served as a hub for this trade, so WildAid has spent time there, hoping to disrupt the shipping channels that connect fishermen to distant markets. Activists approached major airlines and air freight shippers, requesting that these firms ban further traffic in shark fins. By the summer of 2016, several firms had agreed to a ban, including UPS, DHL, and thirty-six international airlines. But FedEx held out, refusing to declare an official moratorium and instead issuing a press release stating that the firm was "opposed to the trafficking of animal parts that were obtained from the exploitation of any species by law."[45]

WildAid activists saw this as a cop-out. They picketed in front of one of FedEx's Hong Kong offices, carrying posters of bloody definned sharks. Explaining WildAid's strategy, activist Alex Hofford told reporters, "What we want to do is cut the supply chain" because "logistics firms like FedEx provide critical links in a long supply chain from illegal boats . . . to the mouths of . . . consumers."[46] Just how critical FedEx was to the fin trade was not entirely clear. There were no publicly available numbers or statistics that showed how many fins FedEx shipped. And as Reuters reported when it covered this story in 2016, roughly 92 percent of shark fin imports into Hong Kong came by boat, not by plane.[47] In other words, cutting the FedEx supply chain would not stop the severing of fins from shark bodies.

Yet the WildAid campaign, like the opioid investigations that occurred at the same time, revealed that FedEx and other air freight shippers had become prime targets in public-health and pollution-control campaigns involving internationally traded commodities precisely because they had become such important channels in the global economy. This was environmentalism and consumer protection that targeted the logistics firms in the global economy, a type of environmentalism that differed from protest campaigns of earlier decades that focused on specific polluting factories. Government regulators and environmental activists saw FedEx as a conduit of capitalism that had the power to close trade networks if it wanted to. According to this paradigm of pollution prevention, for FedEx to be considered a responsible corporate citizen, it had to actively screen out customers who produced wares harmful to people and the planet.

And yet, when FedEx spoke about corporate responsibility in its annual reports, the company offered few specifics about *what* it was shipping around the world; instead, it tended to focus on *how* it planned to ship goods. In the same vein as Coca-Cola and Delta, FedEx's corporate responsibility team promoted "efficiency" as the key watchword in its environmental statements. Its 2017 Global Citizenship Report, for example, highlighted FedEx's efforts to improve packaging, remodel its offices, make its planes more fuel efficient, and green its trucking fleet. There was no mention of the environmental profiles of goods that passed through its corporate system. The firm's sustainability team did note that it had begun monitoring the environmental footprint of "suppliers" it chose to do business with. But the sustainability report made clear that "suppliers" in this case referred to a very limited number of businesses—"equipment manufacturers, fuel companies and transportation service providers"—not the majority of the firms that used the FedEx system to ship goods around the world.[48]

To measure progress on reducing FedEx's carbon footprint, sustainability officers at the firm relied on the metric of "carbon emissions intensity," a phrase that was becoming more popular in corporate sustainability circles by the end of the 2010s. Carbon emissions intensity was essentially a ratio of total greenhouse gas emissions to total revenues generated by the firm. While FedEx self-reported that it had reduced its carbon emissions intensity ratio by roughly 45 percent between 2009 and 2021, the total amount of greenhouse gas emissions generated by the firm during this period went up, not down. Revenues surged dramatically as FedEx's business expanded during these twelve years, meaning the firm could increase its greenhouse gas emissions and still post a carbon emissions intensity ratio that made it seem as if the company were radically reducing its overall carbon footprint. The firm mentioned its total greenhouse gas emissions once in its 2022 sustainability report but focused on its "emissions intensity" improvements at least eight times. It's a clear example of how popular metrics of efficiency, such as carbon emissions intensity, masked the reality that FedEx's greenhouse gas footprint was growing, not rapidly receding.[49]

FedEx's sustainability reports also lacked in-depth discussion of FedEx's worldwide customers. There were few specifics on the firm's thinking about which goods it agreed to ship around the world and which it did not. And yet it was clear that FedEx executives recognized that they could serve an important role as gatekeepers in the economy, keeping harmful products out of consumer markets if they wanted to. Firm representatives stated this directly in the mid-2000s, when FedEx was taking heat from critics and the Justice Department, which alleged the business was not being proactive in preventing online pharmacies from illegally shipping certain controlled substances through its system. FedEx's corporate vice president Bruce Townsend told members of the House of Representatives that FedEx welcomed "the opportunity to continue to work with Congress, law enforcement and our private sector colleagues to find ways to disrupt these networks that negatively impact the health and safety of our citizens."[50] If this was true for narcotics and drugs, why wouldn't it be true for other products known to be harmful to citizens or the environment?

Here, the firm could look to Walmart for inspiration. Back in 2008, Walmart sustainability officers had recognized that they had to look deeper into the companies that produced the goods Walmart sold to its customers if they wanted to create a greener firm—in other words, considering the full life cycle of the goods that flowed through its worldwide logistics system. The Arkansas firm ultimately launched its THESIS Index to measure the environmental profiles of products on its shelves, but as of 2022, Walmart has not been able to realize the goal of having a sustainability score for all the goods sold in its stores. But Walmart's move contained the seed of an idea that could clearly reshape the way express shippers like FedEx think about environmental stewardship in the years ahead.

FedEx's sustainability team would be wise to consider Walmart's approach, because it is clear from the history offered here that firms like

FedEx are increasingly becoming the target of activists and regulators who realize that logistics firms now have outsized power to shape our economy and thus a responsibility to think more boldly about their social and environmental sustainability objectives.

FedEx and its competitors in the logistics space share the same fate as major financial firms in the United States that also became the targets of environmentalists and regulators who believed that by focusing on the conduits of capitalism, they could change the world. A southern financial firm had to learn this lesson the hard way when a group of young activists came to Charlotte, North Carolina, with climbing equipment and hard hats, ready to make a ruckus at the headquarters of one of the largest banks in the world.

PART FIVE

WorldPoints Rewards

Chapter Nine

BANKS GENERALLY DON'T POLLUTE

"You're an idiot if you don't think we have global warming and you don't think we're contributing to it," said Hugh McColl, former CEO and Chairman of Bank of America. Gray-haired and eighty-four, McColl sat in his office in the Bank of America Corporate Center, his desk overlooking the Charlotte skyline strewn with fragmentation grenades that served as paperweights.[1]

That day in 2019, McColl was reflecting on what he had done to promote environmental sustainability when he headed Bank of America from 1998 to 2001. McColl no longer worked for the bank he helped build, but he remained an adviser to the firm's executives. The grenades were obviously meant to remind visitors that McColl was a hard-driving Marine veteran, so often mentioned in business magazines and *Wall Street*

Journal stories detailing McColl's masterful negotiation that in 1998 led to the megamerger between North Carolina's NationsBank and California's Bank of America. But if McColl had all the attributes of a hard-nosed businessman, he also was an idealist concerned about humanitarian causes, including ecological issues. McColl was no fan of government regulation over what he could do with people's money when he oversaw his bank, but the environment was another matter. "What blows my mind," McColl said, "is how anybody can think it's good to roll back regulations on water and air." "Make America Great Again," he laughed. "We're not going to be able to breathe."[2]

Yet during our interview, McColl struggled to identify any major ecological initiatives he had championed during his tenure as CEO and chairman of Bank of America. The bank's annual reports issued between 1998 and 2000 made no mention of the environment or sustainability. McColl noted that this was in part because there were no industry-wide rules rewarding good actors in the banking industry when he was CEO. He said that if his bank had refused to finance a polluting industry, another financial firm would have come along and snagged that business.[3]

McColl's lack of action mirrored that of his predecessors. In the 1970s, when the modern environmental movement was taking off, the president of what would later become NationsBank, Thomas Storrs—McColl's mentor—said he was deeply "concerned" about "air and water pollution" but held that "ecological activities" were not "totally relevant and appropriate to the banking business." In the end, the best thing a bank could do to help the environment was "make a profit." "We can do our own thing as bankers," he claimed, "and do it without having to apologize because we don't have any polluted air or water of our own to clean up."[4]

But the story was much more complicated than Storrs suggested. As it grew into one of world's largest financial firms in the twentieth century, Bank of America and its predecessors accumulated capital by branching aggressively into rural communities that benefited from federal irrigation projects. The bank used that wealth reaped from the countryside to finance suburban sprawl that further increased its financial power. Over time, it channeled its immense assets, which grew exponentially because of interstate mergers and global expansion in the deregulatory banking climate of the 1980s and 1990s, toward international fossil fuel investments, becoming one of the largest financiers of oil and gas enterprises in the world. In the early 2000s, these massive investments caught the attention of environmentalists, especially members of the Rainforest Action Network, who started targeting Bank of America and other commercial banks in their climate change campaigns. A new era in American environmentalism had arrived, one in which activists and lawmakers centered their efforts on cutting off streams of capital before they flowed to polluting industries in the first place. They were trying to find a way to create a more sustainable economy by focusing attention on the cash stacks that financed the smokestacks changing our planet.[5]

The story begins in the mid-1880s, when Hugh McColl's great-grandfather, Confederate veteran and North Carolinian Duncan Donald "D. D." McColl, opened the Bank of Marlboro in the rural border town of Bennettsville, South Carolina. Deposits for the Marlboro bank came exclusively from southern cotton farmers, which put D. D. McColl in a precarious position. "With no other industry in the county bringing in money," he explained to shareholders in the 1880s, "our deposit account is bound to run low in Summer." There was simply "no manufacturing or other producer of money except as fall season by cotton."[6]

Nevertheless, McColl managed to make money, in part because he invested in mills that helped turn cotton into commodities ready for market. In the mid-1890s, during a textile boom that swept through the Gilded Age Carolinas, the Bank of Marlboro financed Bennettsville's first big post–Civil War cotton mill and invested in railroads that connected suppliers with buyers. McColl soon became president of South Carolina's banking association, and in 1911 the Bank of Marlboro reported net profits of over $22,000.[7] That year, D. D. McColl died, leaving his business interests in the hands of his son Hugh Leon McColl (Hugh McColl's grandfather), who, along with his brother David Kenneth "D. K." McColl, expanded the family's influence in the cotton industry over the next two decades.[8]

While the Bank of Marlboro was providing the capital that drew cotton from the land, North Carolina's American Trust Company—formerly the Southern States Trust Company, founded in Charlotte in 1901—was spreading out across the rural countryside, funneling funds drawn from deposits into North Carolina's textile industry. The bank's logo made clear the firm's connection to cotton: it featured a woman drawing yarn through a spinning wheel.[9]

In 1902, the bank opened its first branch outside Charlotte in the rural town of Davidson, free to do so because of North Carolina's lax rules regarding branch banking. This was really one of the keys to explaining why North Carolina became a banking center in the twentieth century. During the Progressive Era and into the Great Depression, many states in the country passed rules limiting statewide branching, especially states like Illinois and Georgia, where smaller banks in less-urban areas feared predation from potential competitors in big cities such as Chicago and Atlanta. In 1916, North Carolina was one of only twelve states that allowed expansive branch banking. Thomas Hills, former executive in North Carolina–based Wachovia (now Wells Fargo), explained the logic: "There was never the big city against the agrarian dynamic" in North Carolina.[10] Though Charlotte and Raleigh today are among the nation's fastest-growing urban centers, for most of the twentieth century the state's Piedmont was structured around a crescent of large towns and small cities. There was no central player in a major metropole that could gobble up small-town banks, and farmers welcomed a branch banking system that could channel capital down country roads to rural communities. Banks, in turn, gained valuable experience in managing operations over large territories, a critical asset that gave the state's financial institutions key advantages when interstate banking began decades later.[11]

In its infancy, however, American Trust gained power not by branching but by becoming a leader in correspondent banking, an arrangement that allowed the bank to extend credit to, offer financial services to, and handle deposits for other banks, including ones operating across state lines. This was important because the McFadden Act of 1927 established rules that essentially prevented banks from opening branches in multiple states. Correspondent banking, therefore, became a method for drawing in capital beyond state borders without violating federal law.[12]

The Bank of Marlboro of South Carolina was one such firm that used American Trust as a correspondent bank. Hugh Leon McColl had seen the Bank of Marlboro through some tough times in the 1920s, when the boll weevil was damaging cotton crops in the American South. But he had done well enough to put his son, Hugh L. McColl Jr. (the father of Bank of America's Hugh McColl), through the University of North Carolina. In 1927, father and son partnered in management of the bank, and seemed to weather the early storms brought on by the stock market crash. In 1931, for example, the American Trust Company gave the bank a glowing review: "You have a distinction that is unique, in that you have the strongest bank in the world, with far more cash on hand and in banks than you have on deposit. The Bank of England and the Federal Reserve cannot match you in strength."[13] Clearly this was hyperbole, but at base, the bank's financials seemed solid. However, that year Hugh Sr. died, leaving Hugh Jr. to deal with the Great Depression, which led him to liquidate the Bank of Marlboro in the 1930s.[14]

After McColl sold the bank, cotton remained king in the family household, as Hugh Jr. joined up with his uncle D. K. to run, among other things, the family's cotton mills. That is when the third Hugh McColl, the future chairman and CEO of Bank of America, was born. It was 1935, the heart of the Great Depression, and the family's cotton business had been hit hard—though not as hard as the Black laborers who toiled in the mills or the sharecroppers who had lost everything in these hard times. Across the state, millworkers' wages had plummeted, and in 1937, Bennettsville-based McColl employees initiated a strike to demand a living wage. This was risky in a Jim Crow town that still rang bells at dusk to signal to Black residents that they had to leave downtown corridors. The family elders decided to shut the mills down rather than work with a labor force that was being unionized by a textile committee of the Congress of Industrial Organizations. In the years ahead, Hugh McColl Jr. decided to focus his energies on the cotton merchant business. After all, there was a lot more money in financing cotton than in growing it or processing it. The younger McColl would learn this lesson well.[15]

On the other side of the country, in the sun-drenched state of California, Bank of America founder Amadeo Peter "A. P." Giannini was following the same country roads to financial prosperity. Born in 1870 to Italian immigrants, A. P. and his two younger brothers grew up on a forty-acre fruit and vegetable farm just outside Alviso, a small port town north of San Jose. Though he later said he "didn't care very much for farming," agriculture was a part of his life from the very beginning. Giannini's birth coincided with a transcontinental railroad revolution that sparked an agricultural boom in the state's productive river valleys. New refrigerated railcars soon allowed California farmers to ditch low-priced commodity crops like wheat in favor of high-dollar fruits and vegetables few farmers could grow farther east. The real boom years began in the 1880s, but Giannini was getting a first-class education in this new big-money business picking apricots and strawberries on his family farm in the 1870s.[16]

The Giannini family was rocked by tragedy in 1876 when an angry worker seeking better pay shot and killed A. P.'s father, leaving Virginia, only twenty-two years of age at the time, to take care of her six-year-old, a toddler, and a newborn baby. The Gianninis muscled through, however, and in 1880 Virginia married Lorenzo Scatena, who became a beloved adopted father to her three boys.[17]

By 1882, Scatena moved the Gianninis to San Francisco, but the family continued to earn a living from the land—just not directly. Scatena joined A. Galli and Sons, a fruit and vegetable commission business, realizing there was a lot of money to be made in selling produce rather than cultivating it. Soon he started his own firm, L. Scatena & Company, and a young A. P., now a teenager, became enthralled with his stepfather's work, traveling with him at night to the bay docks, where he watched the haggling and hustling of the commission men by moonlight. He was hooked, and by fifteen he had dropped out of school so that he could work alongside Scatena.[18]

Giannini often visited farms that other fruit merchants believed were too remote.[19] He stayed close to rivers, because state-sponsored water irrigation projects—dams, canals, aqueducts—were just beginning to turn California's more arid lands into a hydraulic empire. But he journeyed as far as he could go, down the Sacramento and San Joaquin River Valleys and up north into the tangled grapevines of Napa in pursuit of the state's finest fruits and vegetables. And by the end of the 1880s, he was doing business in the sun-kissed orange groves of the Santa Clara River Valley just outside Los Angeles. It was truly an astounding amount of territory for any salesman to cover, and it made the family a lot of money. Scatena's firm became the largest commission house of its kind in San Francisco, bringing in hundreds of thousands of dollars annually. A. P. Giannini, now married to Clorinda Agnes Cuneo, moved into a handsome home in the Italian neighborhood of North Beach.[20]

Flush with cash, father and son started investing in real estate. Here were the seeds of a financial empire that would one day stretch across the globe. When Clorinda Giannini's father passed away in 1902, A. P. Giannini, now thirty-two, joined the Columbus Savings & Loan Society, a firm his father-in-law had helped manage; it had become one of the most trusted banks in San Francisco's Italian American community. As a young director, he immediately got into heated tussles with veteran leadership. At issue were the bank's lending policies, which A. P. Giannini believed were too stingy. He was convinced that Columbus could make more money if it began offering small loans at relatively low interest rates to average, working-class folks in San Francisco and the surrounding countryside. And he pushed the bank to focus on other immigrant communities besides Italian Americans.[21]

At this time, it was a radical proposal; most banks avoided small loans to consumers at all costs, especially people bank lenders did not know well, in large part because it was hard to ensure repayment. As a result, most working-class Americans had to turn to loan sharks, who charged exorbitant interest rates (sometimes as high as 300 percent).[22]

The bank board was not moved by the proposal, so Giannini decided to go out on his own, partnering once again with stepfather Scatena to create the Bank of Italy in 1904. The bank's name was a bit of a misnomer: in addition to Italian Americans, the bank actively sought out people whose families hailed from all over the world, and it even created multilingual divisions to cater to various immigrant communities.[23]

From the very beginning, Giannini bet his financial future on rural America. He immediately went back to the vegetable and fruit dealers he had worked with for many years, offering them small loans—sometimes as small as twenty-five dollars—that no one else would grant. It was all about the little man for Giannini—fishermen, farmers, barbers, and bakers—and in due time Giannini began to build a substantial deposit base, over $100,000 by December 1904. It seemed Giannini had bet right on this small-loan business.[24]

Then, disaster struck. On April 18, 1906, the ground began to shake as tectonic plates shifted below. Buildings crumbled and fires erupted throughout the city, including in North Beach, where Giannini worked feverishly to load his clients' deposits in orange crates packed in a produce wagon. The plan was to rush the bank's money and gold reserves to Giannini's new home outside the city in San Mateo, a feat he just barely pulled off under cover of night.[25]

The cash in the wagon allegedly "smelled like oranges" for weeks thereafter, which was fitting because much of the bank's money ultimately came from produce grown in the countryside.[26] Despite being sullied by the scent of citrus, this cash was legal tender, and Giannini was intent on getting it back in circulation immediately after the devastating earthquake. In a bold move that helped establish him as a legendary banker in San Francisco for decades to come, Giannini set up shop atop a wooden plank positioned across two barrels on Washington Street, down at San Francisco's wharf, and immediately began lending out cash to city citizens who were desperate to rebuild. Many San Franciscans thereafter remained loyal to the bank, moved by how Giannini had responded in the city's darkest hour.[27]

But this produce trader turned banker had bigger plans. As he looked to the future, he felt that the path to profits started on the dirt roads in California's farmland, roads he knew well, since he had first started rambling on them as a young salesman so many years ago. And Giannini was fortunate because, like North Carolina, California approved a state banking law in 1909 that permitted lax branch banking rules.[28]

Giannini had become enthralled by branch banking ever since he heard a 1908 speech by former treasury secretary Lyman J. Gage about its success in Canada. Soon thereafter, Princeton president Woodrow Wilson gave a provocative talk at the American Bankers Association meeting in Denver heralding the financial prosperity that branch banking could bring to America's heartland: "If a system of branch banks . . . could be established which would put the resources of the rich banks of the country at the disposal of whole countrysides [*sic*] to whose merchants and farmers only a restricted and local credit is now open, the attitude of plain men everywhere towards the banks and banking would be changed utterly within less than a generation."[29] Giannini agreed with this assessment, and when he returned from the conference to California, he set himself to creating a branch system in the rural countryside of his home state.

He started near his hometown, San Jose, opening the first Bank of Italy branch outside San Francisco to offer financing to small farmers.[30] Giannini, now in his forties, focused on rural growers, creating a Country Loan Department in 1917 and partnering with the newly created Federal Farm Loan Board established under the Federal Farm Loan Act to help rural citizens gain access to credit. In time, Giannini utilized this law to become a major player in developing Federal Land Banks that were privately owned but backed by federal government guarantees. Historian Judge Earl Glock has argued that these Federal Land Banks, which focused on financing farm mortgages, were the rural roots of a government-backed financial system driven by semipublic entities. As Glock points out, the Home Owners' Loan Corporation, Federal Housing Administration (FHA), the Federal Deposit Insurance Corporation (FDIC), and Federal National Mortgage Association (Fannie Mae), all of which emerged later in the 1930s, were in many ways heirs of the land bank system. In so many ways, then, it was in the countryside that certain rules of our modern mortgage system, which relies heavily on semipublic institutions, were worked out, and Bank of America was right in the middle of the action.[31]

By 1918, the Bank of Italy was the fourth-largest bank in California, with twenty-eight branches statewide.[32] The timing had been perfect. The United States had gone to war, which caused agricultural prices for basic foodstuffs to soar. Approximately 60 percent of the Bank of Italy's deposit growth during the war had come from agricultural communities that received more than 50 percent of the $75 million in loans the Bank of Italy had doled out in 1919.[33]

And growth continued into the 1920s. While some states faced financial woes when commodity prices for wheat and other staple crops plummeted after the war, California hung on, in part because of its diversified agricultural portfolio. Giannini offered mortgages to grape growers in Santa Clara, loans to dairy farmers across the state, and financial services to cotton cultivators in the Imperial and San Joaquin Valleys. As McColl's Bank of Marlboro drew in cash from cotton in the American South, Giannini was doing the same thing in California. His bank reportedly financed half of the state's cotton crop in 1929.[34]

A year later, Giannini merged the Bank of Italy with several banks he had acquired since the 1910s. He called the new enterprise the Bank of America National Trust and Savings Association. The patriotic name had come from a Los Angeles competitor Giannini had bought out who had used the moniker to distinguish his bank from Giannini's foreign-sounding firm. The choice laid bare Giannini's intentions: he hoped to move beyond California and conquer the country.[35]

But that would take time. For now, it was clear Giannini owed his success to the California countryside. Capital derived from cotton, cows, and citrus made his bank rich. By 1930, Bank of America was the largest bank west of Chicago and the third-largest bank in the country; only New York–based National City and Chase held more assets. Bank of America was fast approaching $1 billion in total deposits.[36]

In the 1930s, '40s, and '50s, Bank of America took much of that capital and redeployed it toward suburban development projects insured by New Deal programs that radically reshaped California's environment. Between 1934 and 1950, Bank of America financed more FHA homes than any other firm in the country.[37] With Bank of America money, orchards became lawns. Between 1945 and 1950, bulldozers turned 19 percent of Santa Clara County's pear, peach, apricot, cherry, prune, and walnut lands into suburban developments.[38] By offering automobile financing, the firm helped finance the car culture that made this distended suburban landscape accessible for middle-class white Americans. This was unique because most commercial banks were slow to enter the car loan business, retaining their bias against consumer lending. But Giannini, a pioneer in the business, bucked this trend; by 1949, his bank offered more car loans than any other lender in the state, including General Motors Finance Corporation, an industry leader. The firm also benefited from the GI bill, managing an estimated 10 percent of all home loans in the country issued through that program in 1949.[39] Financing suburbia made a lot of money for Bank of America. By the end of the 1940s, when A. P. Giannini died, Bank of America was the largest bank in the world, with 525 branches in California and $6 billion in assets.[40]

At first glance, Charlotte's American Trust Company looked nothing like San Francisco's Bank of America at midcentury. In 1950, the biggest bank in North Carolina was Winston-Salem-based Wachovia, a firm firmly rooted in the tobacco empires of the state, but even that bank was orders of magnitude smaller than the nation's financial leaders. No North Carolina bank made it on the list of the top fifty banks in the country in 1950.[41]

Yet American Trust Company had grown to become a powerful correspondent bank in the state of North Carolina by following a playbook similar to the one Giannini had used out west.[42] American Trust had drawn heavily on rural America, conducting correspondent business amid the cotton and tobacco fields of the Carolinas. In 1957 American Trust merged with Charlotte's Commercial National to become American Commercial bank.[43] Two years later, Marine veteran Hugh McColl, just twenty-two years old, joined the bank that had long helped his father's cotton business in South Carolina.[44]

McColl believed his connection to rural America helped him in his banking work. When he was at the University of North Carolina, he wrote home to his father, saying how much he had learned about finance watching the cotton market over the years: "The experience of keeping the farm books," he said, "sure helped a lot in accounting." He added, "Maybe you will be able to make a cotton man out of me someday."[45]

In many ways, McColl did become a cotton man. As a young banker, McColl quickly worked up the ranks as a loan agent, traveling to his native South Carolina drumming up correspondent business on cotton farms in his old backyard.[46] Much like Giannini, he worked fast and loose, driving down country roads in an un-air-conditioned Volkswagen and offering farmers loans, sometimes on the very day they asked for them.[47] McColl claimed he could size up a cotton farm in no time. It all went back to his roots. "When I was a little boy, my father made me walk every farm in the county," he said. "He would make me know who owned what land and what it was good for, what it was worth." Just eyeballing, he could figure out how much a particular patch of soil could earn from cotton cultivation. This was country capitalism.[48]

Baltimore native and Johns Hopkins alum Addison Reese ran American Commercial during McColl's early years with the bank.[49] In 1960, taking full advantage of North Carolina's lax regulations, he initiated a merger with Greensboro-based Security National, a bank that had also turned cotton into capital by investing in the region's textile industry. Security had considerable sway in negotiations, holding a strong financial position, but in the end Reese had the trump card: his bank was located in Charlotte, a city where the Federal Reserve had established a branch in 1927. The Queen City would be the headquarters of the newly merged conglomerate: North Carolina National Bank (NCNB).[50] The name said it all: Reese intended to spread out across the entire state, setting up branches that could draw deposits from the farms of the Piedmont, the beaches of the Carolina coast, and the mountains of Appalachia.

The end of the 1960s were boom years for NCNB in part because of changes brought on by civil rights activism. In the summer of 1960, Mayor James Smith agreed to integrate lunch counters (but not other public accommodations).[51] A few years later, Charlotte moved to desegregate all other public accommodations, a landmark event McColl saw as instrumental in helping to make the city an attractive place for business investment.[52] "Maintaining segregation was very expensive and got nothing for nobody," McColl said years later. By the end of the decade, NCNB had appointed a Black man to be president of one of its banks, and the firm's HR department had gotten serious about hiring more minority employees and women. Progress was slow: NCNB reported in 1976 that "minority officials and managers" still represented only 2.6 percent of the bank, and a confidential memorandum at the time concluded that "the *results* of our affirmative action to employ and upgrade minorities and females have been inconsistent with our *efforts* and not in keeping with our desires."[53] Still, the bank happily reported in 1972 that the "dollar value of our loans to minority business enterprises has more than doubled in the past two years."[54]

Black migration out of the state slowed in the 1960s, a pattern that mirrored what was happening throughout the American South as the civil rights movement won advancements for African Americans. (In those years, Black out-migration from southern states decreased by 300,000 people compared to the previous decade.) By the end of the 1970s, more Black citizens were migrating to the American South than leaving. White migration into North Carolina also picked up, in part because of federal investments in the state that went toward social services and Cold War military installations.[55] Hugh McColl put it well: "We were being bailed out by guns and butter; we were being bailed out by the Great Society under LBJ. We had a war going on and [he was] pumping money into . . . welfare and other parts of the state. And so, nobody could lose money in that period."[56] In 1965, as the Vietnam War heated up, NCNB boasted $746 million in deposits.[57]

Deposits available for loan were only part of the magic that brought customers to NCNB in the 1960s. A big game changer, and one that sealed the first partnership between Bank of America and NCNB, was the introduction of credit cards. Bank of America had really revolutionized this industry, becoming the first commercial bank to successfully launch its own card in Fresno, California, in 1958.[58] The BankAmericard quickly spread across the nation, though the American South was slow to catch the fever. In 1966, NCNB became the first bank east of the Mississippi to acquire a license to issue BankAmericards to its customers, and by 1971, it was still the only BankAmericard licensee in the state.[59] The pioneering position helped NCNB pitch itself as a bank primed for national growth, as its customers could now use their cards and expanded consumer credit to make purchases from coast to coast.

In the same way that Bank of America used farmers' money in California to transform rural landscapes in the Golden State, so too did NCNB channel funds earned from an agricultural economy to change North Carolina's ecosystems in the 1960s and 1970s. NCNB loans to Duke Power Company, the utilities firm that dammed many of the state's rivers for electrical power, helped build resort towns along lakes in the Piedmont. Farther east, the bank financed beach homes on the Carolina coast. And in metropolitan areas like Charlotte, NCNB provided the mortgages for suburban homes that overtook fields and farms now more accessible due to the interstate highway system.[60]

Thomas Storrs, who became president of NCNB in 1969, expressed concern that all this bank-financed growth boded ill for the state's environment. In line with the emerging environmentalism of the era, in a 1964 speech Storrs said, "We are rapidly approaching the poison point in both air and water," noting that "automobile wastes alone account for about 50 percent of all air pollution." The car-centered landscape NCNB was financing through its consumer loans was clearly a problem, but in the end, Storrs was not convinced that the bank could solve the nation's waste woes. He believed that "certain types of businesses" other than banks "are more interested . . . and able to participate in" cleanup projects, singling out paper companies and public utilities as key culprits.[61] He continued to discuss environmental issues as the bank entered into the 1970s; in 1972, two years after the first Earth Day and President Nixon's creation of the Environmental Protection Agency and the same year Congress passed the Clean Water Act, Storrs noted that it "has been interesting to watch the banking industry try to climb aboard the ecology bandwagon." He admitted that the nation needed "major surgery," but rejected the notion that banks should be forced to change their practices. "Banks generally don't pollute," he told business school graduates at Rutgers University.[62]

The president of NCNB was taking a clear stand: the company was going to push for economic growth and leave environmental protection to other businesses. But if Storrs liked to portray banks as operating at arm's length from the ecological problems facing the nation, the energy crisis brought on by the OPEC oil embargo of 1973–74 laid bare the fact that banks relied on resource extraction. The energy crisis scared Storrs, who knew that the dwindling of fossil fuel reserves represented an existential threat to his firm. He wrote to Nixon's Council on Environmental Quality (CEQ), urging the administration to approve controversial offshore drilling permits for the outer continental shelf in the Atlantic Ocean. "I do not think the dangers of environmental damage are so great that an outright prohibition of offshore drilling in the Atlantic is justified," he told the CEQ. He also said he favored "gasification of coal, shale oil refining, nuclear energy, solar energy, and other alternatives."[63] Storrs's comment foreshadowed a financial future when the bank would become heavily involved in extreme fossil fuel extraction. Other banks had similar ideas.

One of those firms was Bank of America back in California, now under the direction of Tom Clausen. Serving his fifth year as president of the financial firm, Clausen had just created Bank of America's World Banking Division, part of a big plan for global expansion.[64] Clausen's strategy centered on oil. Clausen wanted to become a world leader in petrodollar recycling—a process that involved taking in deposits in fossil fuel–flush regions of the world (especially the Middle East) and lending that money out to oil-starved developing countries, who would then use those dollars to buy more petroleum to fuel growth in their economies. There was a lot of money in this circular system at a time when oil prices were climbing. And so began Bank of America's fossil fuel frenzy. By the mid-1970s, Clausen moved tremendous amounts of capital through the petrodollar recycling machine, and some of this money ended up financing new fossil fuel exploration around the world. Enticed by soaring oil prices, the firm offered substantial financing to oil exploration and drilling companies. Bank of America's loan portfolio included investments in the oil fields of Alaska, Brazil, Bolivia, Mexico, and beyond.[65]

California's Bank of America was not the only financial institution developing a deep dependency on the oil industry at this time. Banks in the oil-rich states of Texas and Oklahoma also invested heavily in oil during the boom years of the 1970s. Firms such as First Republic, based in Dallas, rushed out real estate deals and loans to oil services companies in a mad dash to make as much money as possible as oil prices rose to record highs. But the flood of new oil production soon created a glut that brought prices spiraling down in the mid-1980s. For banks like First Republic, the consequences were devastating.[66]

In this moment, Charlotte banker Hugh McColl thought he could exploit the chaos in the oil market to vastly expand North Carolina National Bank. McColl had worked his way up the ranks of NCNB to become bank president in 1974 and then chairman and CEO in 1983, when NCNB was fast becoming one of the most powerful banks outside of New York and California. In the early 1980s, McColl lobbied state legislatures to approve what became known as the "Southeastern Compact," a regional agreement among southern states east of the Mississippi to allow interstate branching within the region. NCNB got what it wanted when a 1985 U.S. Supreme Court decision ratified the compact. A fiery McColl moved immediately to take over other banks outside North Carolina. Meanwhile, Wachovia—for many years North Carolina's bank leader—dragged its feet, losing ground year by year to their in-state archrival.[67]

This was the situation when the oil market collapsed in the mid-1980s and First Republic of Dallas started drowning in bad loans.[68] The FDIC was desperate to find some way to save this massive Texas bank, fearing that a bankruptcy would send shock waves through the state's economy. McColl seized on the opportunity and approached the FDIC about buying First Republic. The federal agency approved the takeover in July 1988 and assumed a portion of First Republic's liabilities associated with bad loans. In the end, the government never forced NCNB to repay the full debt accrued from unwise fossil fuel investments.[69]

A party ensued. In Dallas, NCNB employees downed "flaming tequila drinks," and back home in Charlotte, Jane McColl, Hugh's wife, sipped on a massive bottle of champagne that filled an entire bathtub. It was the "turning point of the company," McColl said years later. "After that, we did think we were going to build the biggest bank in the country."[70]

This acquisition tied NCNB more closely to the oil business than ever. In 1990, NCNB's Dallas office reported that the firm's fossil fuel investments had increased by nearly a quarter in just the first six months of the year. Two years later, NCNB became NationsBank.[71]

In 1994, good news came from Washington. Congress passed the Riegle-Neal Interstate Banking and Branching Efficiency Act, which officially repealed restrictions on interstate banking. Like Fred Smith of FedEx, McColl had courted the Clinton administration, requesting help in getting this law passed. When the bill became a law, McColl moved to gobble up banks in the Midwest and beyond. McColl boasted that his corporation was now the largest bank in the nation, and one that had become, by 1996, "one of the largest oil and gas lenders in the world."[72]

Chapter Ten

BANK OF COAL

As Hugh McColl spread his bank west in the early 1990s, Bank of America in California had just emerged from a series of nightmarish events that had threatened its solvency. The petrodollar recycling scheme Tom Clausen had put in place in the 1970s turned sour by the early 1980s, when Latin American countries began to default on their high-interest loans.[1] However, Latin American debt was only part of the problem. In California, the bank's extensive investments in rural communities became huge liabilities for the firm. In 1985, Bank of America remained the biggest "farm bank" in the United States, with millions of dollars in loans outstanding to California growers. Many of these loans underperformed. Farm real estate prices were falling alongside commodity prices as production overseas ate into U.S. farmers' market share. Interest rates were rising, placing heavy burdens on farm families, and Jimmy Carter's 1980 Soviet grain embargo meant farmers had lost key export channels. These factors together produced the 1980s farm crisis, which affected the Midwest the most heavily, but other farming communities as well. Bank of America posted over $500 million in losses in 1986. Though the company had been the biggest bank in the country in 1977, a decade later it was not even in the top ten.[2]

As the firm sought to improve its financial position, managers in the bank became concerned about the water demands of their farming clients. In 1991, five years into a long California drought, Bank of America began requiring farmers to show how they would meet their water needs for the year before offering a loan. To be clear, press reports revealed that Bank of America managers were largely concerned about water at this time because of economic considerations (farmers' inability to pay back loans) rather than out of a deep concern for the health of California's farming ecosystems.[3]

Nevertheless, the early 1990s marked the first time the bank began to develop anything that resembled a corporate environmental responsibility program. At this time, many corporations, not just banks, were beginning to launch environmental-sustainability programs because of new environmental regulations as well as renewed environmental activism following the 1984 Bhopal gas leak disaster at a Union Carbide plant in India, which killed thousands of people, and the tragic spectacle of the 1989 *Exxon Valdez* oil spill. The Union Carbide pesticide plant leaked a toxic gas called methyl isocyanate into poor communities around the facility, and government officials now claim that roughly 15,000 deaths can be attributed to chemical exposure from this accident. The press dubbed it "the world's worst industrial disaster." It occurred less than five years before the *Valdez* tragedy, which was the biggest oil spill in U.S. history to that point and affected nearly 1,300 miles of coastline, resulting in the deaths of seals, killer whales, salmon, and tens of thousands of seabirds. The firms promising publicly to make the biggest changes after these headline disasters were not banks but chemical companies and other industrial firms, which felt more intense pressure to adopt eco-friendly practices. Congress had recently passed the Right-to-Know Act of 1986, which required companies to report any toxic chemical releases that might come from their facilities. Because Bank of America and other financial firms did not directly own plants spewing deadly compounds into the water or air, in the early 1990s they were less visible targets both for regulators and for activists.[4]

Nevertheless, by the early 1990s, Bank of America began hiring environmental analysts to assess liabilities the firm might face under the Comprehensive Environmental Response, Compensation, and Liability Act (CERCLA), also known as the Superfund Act. Passed in 1980 and enhanced by amendments in 1986, the Superfund Act worried bankers because a series of federal court cases suggested that banks could be held liable for toxic waste on foreclosed property held by these financial firms. The Asset Conservation, Lender Liability, and Deposit Insurance Protection Act, passed by Congress in 1996, calmed bankers' fears by limiting lender liability under CERCLA, but despite these assurances, Bank of America continued to rely on environmental analysts when making loans.[5]

In 1991, the firm announced that it had joined the Environmental Protection Agency's Green Lights Program, an initiative to reduce corporate energy demands by installing energy-efficient lighting in company offices. At the time, United Nations member states were planning for the 1992 Conference on Environment and Development, popularly known as the "Earth Summit," where the United States and over 150 other countries signed the first global climate agreement to reduce greenhouse gas emissions, and member states were discussing what financial firms could do to contribute to environmental sustainability. New boutique banks, some established in Europe in the 1970s and 1980s, were channeling their capital resources toward social and environmental sustainability objectives. The historian Geoffrey Jones aptly noted that the combined assets of these "social banks" represented "a tiny part of the global financial system."[6] Still, the idea that banks should be focusing on larger environmental issues was part of the public conversation. Between 1991 and 1998, Bank of America said it had reduced its greenhouse gas emissions by approximately 40 million pounds through the Green Lights Program, which equated to removing some 3,000 cars off America's streets.[7]

But even as it marginally reduced carbon emissions with better light bulbs, it was still fueling tremendous fossil fuel consumption by stimulating suburban sprawl, and this topic became a concern within Bank of America, especially as the firm poured more money into suburban development projects. In 1992, the bank had merged with Los Angeles–based Security Pacific, making it "the largest lender in the United States for office, shopping centers, new housing tracts and other real estate."[8] The ecological impact of Bank of America's specific real estate financing is hard to precisely quantify, but scholars studying the environmental effects of suburban sprawl during this period have made some key findings. Matthew E. Kahn, a Columbia University economics professor, concluded in a 2000 study that suburban residents drove 31 percent more than urban residents and used roughly twice as much land as people living in cities.[9] And suburban development naturally required copious quantities of asphalt, metals, and timber. One study conducted a few years before Bank of America purchased Security Pacific estimated that suburban residential development consumed over a third of all timber harvested each year in the United States.[10]

Bank of America executives realized that their financing of suburbanization was having adverse effects on ecosystems, especially in the bank's home state of California. In 1995, the firm penned a report with state officials and NGOs titled "Beyond Sprawl" that made a passionate appeal for a new era of sustainable development: "Sprawl," the report declared "compromises one of the most essential assets of California—the beauty and drama of its landscape." The publication showed that 95 percent of California's wetlands had been destroyed by suburban development and noted that over 30 percent of the state's air pollution came from cars traveling across sprawling highways. But when asked how the report would practically reshape Bank of America's home loan program, an eco-policy representative at the bank offered little hope for real change. "We will finance housing almost anywhere," he said, "as long as there is a good market for the house."[11]

The bank was equally hesitant to take meaningful steps to address climate change. In 1991, the United Nations Environment Programme inaugurated its Finance Initiative (UNEP FI) to bring together financial firms to combat climate change, and Bank of America soon became a member. But over the course of the next decade, the firm did little beyond initiatives such as the Green Lights Program to deal with global warming, and it did not take focused steps to eliminate fossil fuel investments. Looking back at the firm's early sustainability goals, Bank of America admitted in 2007 that much of its work in the 1990s focused on reducing the amount of paper the company used and making company offices more energy efficient. It was not primarily focused on lending portfolio overhauls.[12]

And the bank's lending portfolio was getting bigger. With the passage of the Riegle-Neal Interstate Banking and Branching Efficiency Act of 1994, Bank of America's capacity to affect climate change expanded as it sprawled out across state borders in an effort to become a coast-to-coast bank. Bank of America was moving east, setting up branches in ten western states and breaking ground on its first East Coast bank in Washington, D.C. In 1995, Bank of America believed a key next step was a NationsBank buyout, which would give it unprecedented geographic reach, but Hugh McColl had other plans.[13]

McColl had watched as the American South's economy boomed just as his bank was freed from interstate banking restrictions. Unlike his California competitors, there is no evidence NationsBank ever paused to focus on the environmental effects of the rapid suburban growth it was helping to finance in the Sunbelt. It was full steam ahead as McColl's bank expanded its business in places like Florida, where ecosystems were rapidly and radically being transformed into suburban homes and resort communities. By 1998, three years after Bank of America had approached NationsBank about a merger, McColl was outpacing Bank of America. The North Carolina firm was now a bigger bank. With the upper hand, McColl began Scotch-infused negotiations in a California hotel room and convinced Bank of America CEO David Coulter to accept certain demands, including that the new bank headquarters would be in Charlotte and that the board would have eleven NationsBank representatives to Bank of America's nine. This was to be a southern company, one that would now have outsized influence on global economies and ecologies. In 1998, when these two banks became one, Hugh McColl was at the helm. The combined assets of the new financial firm totaled more than $600 billion.[14]

Now larger than ever under McColl, Bank of America could turn natural resources mined, farmed, and piped from ecosystems across the globe into tremendous financial capital pools that could be deployed in massive industrial projects, including fossil fuel extraction. In 1999, a year after the merger deal and two years after the Kyoto climate summit, President Clinton repealed provisions of the Glass-Steagall Act that had prevented mergers between commercial and investment banks since the Great Depression. In the years ahead, Bank of America accumulated even more financial strength as it gained new access to securities markets, and so did other financial firms. In 1980, not a single bank made it onto the top twenty of the Fortune 500 list. But three decades later, Bank of America, Citigroup, JP Morgan Chase & Co., and Wells Fargo all did.[15]

By the end of the 1990s, environmental organizations recognized that financial firms had grown incredibly powerful in the era of bank deregulation, and they began campaigns to highlight how big banks were fueling climate change and ecosystem degradation around the world. The Rainforest Action Network (RAN) played a leading role, initiating its first big commercial and investment bank protests in the early 2000s. Founded in 1985 and run by fiery activist Randy Hayes, RAN had attacked the World Bank and multinational firms like Burger King in the past, but it had never adopted a focused and sustained strategy centered on large financial firms. Following a successful 1999 campaign that forced Home Depot to limit its use of old-growth timber, RAN launched its new plan of action. It started with New York–based Citigroup, demanding that the firm end loans supporting logging in old-growth forests, cease funding for oil drilling in environmentally sensitive ecosystems, and begin the process of terminating financing for fossil fuel extraction.[16]

The organization was unrelenting, hanging banners at company headquarters in New York, sending campaigners to chain themselves to the entrances of Citi branch offices across the country, and taking out full-page advertisements in national newspapers, including one that featured a photo of Sanford I. Weill, CEO of Citigroup, and read: "Put a Face on Global Warming and Forest Destruction." "We're not spending all of our time trying to influence legislators and governments anymore," said RAN campaigner and later Sierra Club director Michael Brune about RAN's shifting strategy: "We're going after the root of the problem."[17] In April 2003, three years into the Citi campaign, RAN sponsored a stunning TV commercial featuring actress Susan Sarandon cutting Citi credit cards as a protest against the company's funding of deforestation around the world. Weill ultimately decided to sit down with Hayes to negotiate. In January 2004, Citigroup agreed to halt any financing of illegal timber cutting and said it would scrutinize its fossil fuel investments, although it made no major commitments to end oil and gas financing.[18]

Bank of America soon fell in line. Ken Lewis, who took over from Hugh McColl in 2001, agreed to work with the Rainforest Action Network (RAN). In 2004, Lewis signed an agreement with RAN, pledging to cut Bank of America's carbon emissions (7 percent by 2008) and to end financing of logging in tropical old-growth forests.[19] "At Bank of America," Ken Lewis said, "we know we have an opportunity and responsibility as leaders to promote sustainable, environmentally sound economic growth in all our communities."[20]

As part of the 2004 agreement with RAN, Lewis created an "environmental council" tasked with looking over the bank's asset portfolio and recommending areas where the firm could improve its ecological footprint. The bank also said it would continue to work closely with the Coalition for Environmentally Responsible Economies (CERES), an organization founded in the wake of the 1989 Exxon Valdez oil spill that sought to establish effective metrics for assessing the environmental performance of multinational firms, including banks. In addition, the firm noted that it would remain committed to the United Nations Environment Programme's Finance Initiative, started in 1991 and reconstituted in 1997, which sought to bring global financial firms together to address critical environmental issues. In May 2004, RAN gave Bank of America kudos, running a full-page ad in the *New York Times* about the Charlotte bank's pledge to reduce its emissions. The ad read: "What did Bank of America do for Earth Day? The right thing."[21]

But this period of good feelings between RAN and big banks was short-lived. In 2005, RAN organized a series of demonstrations targeting J. P. Morgan for financing industries that were contributing to climate change. In May, activists visited the Greenwich, Connecticut, neighborhood of the J. P. Morgan Chase CEO William B. Harrison Jr., hanging "Most Wanted" posters that dubbed Harrison the "Billy the Kid" of eco-crime. A month later, Harrison agreed to launch a series of environmental programs designed to address key RAN concerns, such as stopping loans to logging companies operating in sensitive ecosystems.[22]

RAN then refocused its energies on Bank of America. In September 2007, RAN activists put up huge signs outside the headquarters of Merrill Lynch that said Bank of America was accelerating global warming through its fossil fuel investments. Bank of America executives sitting inside the building were attending the annual meeting of the Carbon Disclosure Project (CDP), a British organization started in 2000 that offered carbon emissions monitoring services to financial and industrial firms. Bank of America was a CDP member, but RAN hoped to show that there was a big difference between voluntary emissions disclosures and the hard work of actually divesting from fossil fuels.[23] The following month, RAN activists hung a fifty-foot banner from a construction crane positioned in front of the bank's corporate tower in Charlotte. The banner read: "Funding coal: killing communities."[24]

Facing unrelenting protests covered by the *New York Times*, CNN, and other popular media outlets, in 2007 Bank of America announced a major plan to use its financial resources to deal with global warming. That year, the company made a pledge to spend approximately $20 billion over the course of a decade to "address climate change." Working with CERES, the bank also launched a Brighter Planet Visa card, which offered bank clients the option to commit reward points toward environmental initiatives. In 2009, the firm paid for a full-page ad in the *New York Times* announcing its intentions and happily noted in its annual report that it had already "delivered more than $5.9 billion in lending, investing, and new products and services, including nearly $900 million in financing for renewable and energy efficiency projects in 2009 alone."[25]

All seemed well, but these corporate sustainability statements belied the reality that Bank of America continued to be a prime financier of fossil fuel industries. In its 2010–11 annual report, RAN stated that Bank of America had channeled more than $4.3 billion in just the past two years toward both the extraction and consumption of coal, making it "the largest financier of coal in the country."[26] In 2010, RAN, working with the Sierra Club and BankTrack (an organization started in 2003 to monitor sustainability initiatives in the financial sector), published its first report card detailing major banks' fossil fuel investments. Sierra Club had been fighting to stop coal mining through its Beyond Coal campaign, which began in 2002, but RAN had been the real leader in turning a spotlight on financial firms. The 2010 report, which centered on mountaintop removal, noted that nine banks, including Bank of America, Wells Fargo, and Citi, contributed "more than $3.9 billion in loans and bond underwriting to companies practicing mountaintop removal coal mining."[27]

With this information in hand, RAN redoubled its protest efforts against Bank of America with a national campaign, calling on people to close their accounts with the bank. At the time, RAN was feeding off popular unrest against the banks that had festered since the recession and bailouts of 2008. The same year that the Occupy Wall Street movement surged in New York City, RAN initiated a broad grassroots campaign in Charlotte that involved frequent demonstrations at company headquarters. In 2012, right before Bank of America's annual shareholders' meeting, RAN activists pulled off a daring feat, rappelling down the face of Bank of America Stadium and hanging a seventy-foot banner that dubbed Bank of America the "Bank of Coal."[28]

Two months after the stadium demonstration, Bank of America announced a new pledge to commit $50 billion over ten years toward mitigating climate change, but RAN kept the pressure on.[29] In 2013, the organization infiltrated Bank of America's shareholders' meeting, and twenty-four activists took to the microphone to protest the company's continued investments in fossil fuels. At one moment, the flustered CEO Brian Moynihan fired back from the stage, "Is there anyone out there who has a question that isn't about climate change?"[30] Following the meeting, Bank of America acted, saying it would get out of the mountaintop removal mining business that had devastated Appalachian communities in the American South. In its 2013 annual social responsibility report, the firm also happily reported that it had spent more than "$27 billion in financing for low-carbon activities, such as energy efficiency and renewable energy." That year, the company announced that it would issue $500 million in "green bonds" to "finance energy efficiency and renewable energy projects."[31] Bank of America thereby gained recognition as an environmental leader in financial circles. By 2015 it earned listing on the World and North American Dow Jones Sustainability Indices (started in 1999), which recognized environmentally progressive firms. The Carbon Disclosure Project and the White House also honored the bank among sustainability leaders.[32]

But in 2016, just as Bank of America announced that it had directed $15.9 billion "to support clients connected to clean energy and other environmentally supportive activities," RAN reported that the firm also had spent $36 billion—more than twice as much—financing the fossil fuel industry, making it the third-largest financier of fossil fuel firms in the world. This included extreme operations, such as ultra-deepwater oil drilling, where Bank of America came in third among all other world banks for the period beginning in 2016 and ending in 2018.[33] Fracking was also a significant part of Bank of America's investment plan in the twenty-first century. In 2019, the same year Bank of America announced a new climate pledge to devote $300 billion toward "clean energy finance" by 2030, the bank was one of the top financial firms funding fracking around the globe, spending over $10.9 billion that year alone on various drilling and pipeline deals. To put this number in perspective, Bank of

America spent only $8.9 billion on solar energy projects between 2007 and 2018.[34]

The Rainforest Action Network and other environmental organizations continued to publish "Banking on Climate Change" reports trying to raise awareness about the connections between high finance and rising temperatures.[35] And in 2020, the climate activist Bill McKibben coauthored a *New York Times* op-ed urging readers to take their money out of banks that were spending billions on oil, gas, and coal extraction and consumption. He specifically called out Bank of America in his appeal.[36]

Investment firms were also taking action. In 2017, a group of financial institutions managing nearly $1 trillion in assets fired off letters to the CEOs of Bank of America and other big banks demanding that they take more aggressive steps to address climate change. ShareAction, an organization started in 2005 that brought together equity managers to put pressure on companies contributing to climate change, spearheaded this effort and labeled it the Investor Decarbonisation Initiative.[37] In January 2020, BlackRock, the largest financial asset manager in the world, said it, too, would begin a major initiative to address climate change, channeling the $7 trillion it managed away from firms that failed to meet certain environmental standards.[38]

By the end of the 2010s, environmental campaigns against banks initiated in the early 2000s began to spur political change on Capitol Hill. In April 2019, during a House Financial Services Committee hearing on banking regulations, Representative Rashida Tlaib (D-Mich.) grilled Bank of America's Brian Moynihan and other major bank CEOs on their failure to reduce financing for the oil and gas industries: "You are green-washing your own track record," she said, "and duping the American people into believing that you are helping to address climate change."[39] That same year, Senator Elizabeth Warren (D-Mass.) and her colleague Brian Schatz (D-Hawaii) proposed the Climate Change Financial Risk Act of 2019, which tasked the Federal Reserve with investigating climate change risks in the financial system.[40] When Senator Warren ran for president a few months later, she spoke passionately about the need for climate change amendments to the Dodd-Frank Wall Street Reform and Consumer Protection Act of 2010, a law set up to deal with predatory lending practices and other bad banking behavior that caused the 2008 recession. As Warren pointed out, this law had no teeth when it came to forcing banks to address their contributions to climate change.[41] But she said it could.

Language in that law already empowered the Treasury Department's Financial Stability Oversight Council to investigate "systemic risk" and push big banks to adjust investments that would be harmful to the public interest. As Warren explained, such action awaited an administration willing to put the plan in place.[42] In May 2021, President Biden signaled that his administration was sympathetic to Warren's position, signing an Executive Order on Climate-Related Financial Risk, which tasked the Secretary of Treasury and heads of other federal agencies with, among other things, discussing "approaches to incorporating the consideration of climate-related financial risk into their respective regulatory and supervisory activities."[43] But just how these discussions would affect financial regulation in the United States was still uncertain as this book went to press.

By 2021, Bank of America had clearly entered a new era in American finance. In the face of continued pressure from environmental groups such as RAN and the Sierra Club, as well as from investment coalitions like ShareAction and progressive politicians on Capitol Hill, Bank of America pledged billions of dollars to climate change remediation and became a leader in marketing new financial instruments, such as green bonds, to develop new eco-friendly infrastructure. Bank of America's leadership also said it supported the Paris Climate Accords, that they were committed to reaching "net zero" emissions by 2050, and that they had plans to be more transparent in their emissions reporting.[44] Yet these commitments happened even as the bank continued to make major financial investments in fossil fuel extraction. In the 2000s and 2010s, Bank of America was ultimately reactive, responding to specific RAN and Sierra Club campaigns against mountaintop-removal mining in states such as West Virginia and drilling in Alaska's Arctic National Wildlife Refuge, while never fully committing to a complete divestment from new oil and gas drilling projects. In fact, as this book went to press, the Sierra Club and RAN were actively putting pressure on Bank of America to stop channeling money toward Enbridge, the company building a new "Line 3" pipeline that would expand oil firms' capacity to ship Canadian tar sands oil to U.S. refineries.[45] And so the fight continues, as activists and regulators battle with this Charlotte bank and other financial firms in hopes that by redirecting capital flows away from fossil fuel projects, they can help prevent ocean waters from overtaking the very avenues that run toward Wall Street.

Conclusion

A few years after graduate school, I took a walk in the woods along the Chattahoochee River behind the house where my parents lived in north Atlanta, and I noticed a pipeline I had seen many times before, buried behind some underbrush. As a kid I had always looked inquisitively at this pipeline, briefly pondering why it was there, but never really giving it too much thought. This time, however, I took a closer look. I decided to ramble near to the line, and when I did, I saw a sign that read: COLONIAL PIPELINE COMPANY.

Like many Atlantans, I had no idea that this conduit right in front of me was the largest pipeline for refined fossil fuel in the United States, capable of channeling more than 100 million gallons of fuel a day up and down twelve states, stretching from Houston, Texas, to Linden, New

Jersey. The Colonial Pipeline was an intricate network of 5,500 miles of pipe that provided nearly 45 percent of all the fuel used on the East Coast. Back in 1962, a conglomerate of nine of the biggest petroleum companies in the United States, including Texaco, Gulf, Phillips, Sinclair Mobil, Standard Oil of Indiana, Continental, Pure, and Cities Service, set out to build this pipeline to swiftly move petroleum to markets far from oil infrastructure in Houston. The impetus for construction was a 1960 oil tanker strike that hindered big oil firms' ability to deliver their fuel to customers. Alpharetta, Georgia, a suburb of Atlanta, became the headquarters for the firm, in large part because this was essentially the midway point for the pipeline. It also was a strategic location in the growing Sunbelt South, a place that was fast becoming a major market for fossil fuel companies. The conglomerate called the conduit Colonial because it stretched through many of the thirteen original American colonies.[1]

In most years, Colonial Pipeline Company remained, as my hometown newspaper put it, a "quiet presence" in Atlanta, but there were times when the firm made headlines—and for all the wrong reasons.[2] In the 1990s, there were several incidents where sections of the pipeline burst in Virginia, South Carolina, and elsewhere, creating toxic spills that caused large fish kills and degraded ecosystems. I didn't know it then, but one of these big spills happened in 1998 near the baseball park where I first learned how to throw change-ups.[3] In 2003, the Justice Department forced Colonial Pipeline to pay a $34 million fine to the EPA for failing to prevent several oil spills that resulted in the uncontrolled release of roughly 1.45 million gallons of oil into the environment. This was the "largest civil penalty" any corporation had paid to the EPA at that time.[4]

Sometimes breaches in the pipeline were so catastrophic that they made national news. In 1994, for example, the *Wall Street Journal* reported on a flood in East Texas that forced Colonial to shut off flow for a few days. The newspaper explained that the disruption caused havoc "as fears of further turmoil sent product-prices soaring." In the years ahead, hurricanes and tropical storms, many likely fueled by climatic changes caused by the burning of the very oil that flowed through the Colonial Pipeline, caused the Alpharetta firm to temporarily cut off the stream of fuel to hungry customers up and down the Atlantic seaboard.[5]

And in May 2021, virtually everyone living on the East Coast became familiar with the Colonial Pipeline when a mysterious hacker group called DarkSide used ransomware to break into Colonial's computer system, temporarily forcing the firm to shut down operation. The breach (coupled with "panic buying") led to serious gas shortages in towns and cities on the East Coast. Again, millions of Americans were reminded, however briefly, that their lives were intimately connected to a southern company that managed a critical conduit pulsing the lifeblood of the American economy up from Houston, Texas.[6]

In the immediate days after the 2021 ransomware attack, reporters reached out to several companies threatened by the pipeline disruption. Among them was Delta Air Lines, whose airport headquarters in Atlanta, Georgia, received nearly 70 percent of its jet fuel supplies from the Colonial network. In response to press queries from CNN and other outlets, Delta CEO Ed Bastian said, "We're staying in close contact with the people at Colonial. We're monitoring the situation very carefully. Right now, we're not having any impact at Delta. . . . They're telling us that they expect the supplies to be back on hopefully by the end of the week, and as long as those predictions come true, we'll be OK."[7] As the *Atlanta Journal-Constitution* rightly noted, one reason Delta was able to weather the temporary shortages was that it owned its own oil refinery in Pennsylvania, which helped it fuel its planes farther up the East Coast. But as Bastian intimated, if the Colonial shutdown continued longer than expected, there was going to be real trouble at Delta headquarters in Atlanta.[8]

FedEx was coyer when discussing its position. "Shipping Companies and USPS Won't Say Whether Gas Shortage in Southeast Is Affecting Operations," ran the CNN headline. The news agency reported that it had reached out to a representative of the Memphis-based air express company who said the firm was "communicating frequently with common carriers and mobile fuel vendors to monitor and adapt to potential fuel shortages."[9] The *Charlotte Observer* heard the same exact thing.[10] This made sense. After all, if the Colonial mishap had not been resolved in relatively swift fashion, FedEx's logistics team would have experienced massive headaches trying to fill the tanks of over 200,000 trucks and more than 650 aircraft in its transportation fleet.[11]

Walmart was perhaps most directly affected by the Colonial shutdown. In many places on the East Coast, the retail giant was unable to replenish fuel at its company-owned gas stations sited at its stores. Walmart management began building gas stations at Sam's Clubs in the 1990s, and it initiated a concerted effort to locate fueling stations at its other big-box retail stores around 2016. During the Colonial crisis, Walmart had to turn away customers at several East Coast stores because it simply had no fuel.[12] And like FedEx, the situation could have been much worse if operations at Colonial had not been restored quickly, given that Walmart also depended on thousands of diesel-filled tractor trailers to haul goods to its stores.

The Colonial crisis laid bare the fact that many of the southern companies that ran the high-tech logistics systems, air cargo fleets, and retail satellite technologies channeling goods through the commercial jet streams of the global economy were nevertheless chained to a fossil fuel pipeline that connected them back to an old extractive economy born of the Gilded Age. To be sure, the southern companies featured in this book that created some of the most powerful conduits of commerce in our modern economy have found a way to spread their brands across the globe. Nevertheless, they have remained deeply dependent on powerful fossil fuel firms, many—such as Exxon—based in the American South, whose oil remains the propellant that keeps goods, people, and capital flowing through our economy.

Colonial Pipeline is not the only southern firm that has been the target of international cyberattacks in recent years. In 2012, hackers infected Bank of America's and other financial firms' computer systems with harmful malware that prevented consumers from using the firms' websites. Experts said it was "one of the biggest cyberattacks they've ever seen," and ultimately the breach forced Bank of America CEO Brian Moynihan to invest nearly $1 billion annually to combat ransomware hacks.[13] And in 2017, FedEx was hit hard by the NotPetya cyberattack that many experts suspected came out of Russia. The hack involved the use of "wiper software" that deleted critical data FedEx needed to operate portions of its logistics systems managed by a subsidiary that operated in Ukraine. In the end, FedEx reported that the attack cost the firm somewhere around $400 million.[14]

These hacks made clear that companies managing the conduits of capitalism were now big targets, and not just of cybercriminals. As we have seen, environmentalists have recently recognized that focusing on firms that serve as commercial channels in our economy can be a powerful way to bring about broad transformative change. Environmentalists are no longer simply concerned about smokestacks but are instead turning their attention toward the big retailers, logistics giants, and banks that orchestrate, rather than own, the extractive industries that are remaking our planet and our climate.

A book on the environmental impact of contemporary business can't end without acknowledging the global behemoth that is Amazon, the corporation in second place on the 2021 Fortune 500 list, with roughly $386 billion in revenue. The firm, based in Seattle, Washington, has radically reshaped American commerce, making it easy for consumers to purchase almost anything they want online with just a click of a button. Jeff Bezos, the company's founder, has become rich—so rich, in fact, that he has been able to fund his own private trip to space. But Amazon's gargantuan size has also drawn the attention of labor activists and eco-conscious consumers who believe changing Amazon may well be a means of changing our whole economy.[15]

Amazon is an easy target, but it is important to point out that the number-one spot on the 2021 Fortune 500 list was still held by a once-scrappy company from Bentonville, Arkansas, called Walmart, a firm that arguably perfected much of the sophisticated retail logistics that Jeff Bezos later built on to create his Amazon empire. It may be convenient, in other words, to think of our point-and-click express delivery economy as a product of Seattle tech genius, but the truth is that many of the Amazon packages that arrive at consumers' doors come via an express delivery system that grew out of Memphis, Tennessee, and logistics thinking first pioneered by a guy named Sam Walton.[16]

Jeff Bezos admitted as much. In May 2019, Bezos gave a speech in Washington, D.C., in which he confessed that much of his fortune and fame came from work done by other firms. "I started Amazon in 1994," he said, adding, "All the heavy-lifting infrastructure needed for Amazon was already in place. We did not have to build a transportation system to deliver packages. It existed already. If we'd had to build that, we would have needed billions of dollars in capital. But it was there. It was called U.S. Postal Service, Deutsche Post, the Royal Mail, UPS, and FedEx. We got to stand on top of that infrastructure." Bezos went on: "The same was true of payment systems. Did we have to invent a payment system and roll that out? That would have taken billions of dollars and many decades. But no, it already existed. It was called the credit card." Here was a direct acknowledgment of the role Bank of America had played in facilitating the rise of the Amazon empire. "Infrastructure lets entrepreneurs do amazing things," he concluded.[17]

The central thing we must recognize is that much of the infrastructure Bezos depended on to build his company was refined and honed in the American South, where southern businesses learned how to navigate the rural roads to our Amazon economy long before the internet was invented. The American South may have looked like a backward place for much of the twentieth century, in part because of the archaic and barbaric social institutions it fostered, but the South was also a place where radical changes in commercial logistics were made that enabled companies to move people, goods, and capital more swiftly through the economy than ever before. By figuring out how to service a vast southern countryside, firms like Coca-Cola, Delta, Walmart, FedEx, and Bank of America figured out how to conquer the world.

But that world is now facing a global ecological crisis, in part because of the fast-paced, long-distance commerce many of these firms helped make possible. The 2021 Intergovernmental Panel on Climate Change (IPCC) report made clear that to avert serious climate calamity, nations were going to have to radically reduce greenhouse gas emissions coming from factories, cars, and trucks across the globe over the course of the next decade.[18] Companies like the ones featured in this book, firms that make their money by being conduits of commerce, are really the lungs of our global economy, pulling in goods from one part of the world before releasing them to another. These regulators of commodity flow deserve our attention in this critical moment of climate peril. Southern firms' quest to service the countryside, noble as that quest may have been at its origins, led us down rural roads to an economy that now devours tremendous ecological and energy resources to satisfy the instantaneous gratification of consumer wants. By putting pressures on these firms to do more than just make their buildings more efficient and their trucks more eco-friendly, by asking these firms to make hard choices about whether they should ship certain goods in the first place, we can make rapid and systemic changes to our global economy.

But if we wait for these firms to adjust to the climate crisis on their own, it may be too late. Only in the face of continued agitation from tireless activists did Bank of America agree not to support for drilling in the Arctic National Wildlife Refuge or to cease loaning money to coal firms that were laying waste to 300-million-year-old mountains in Appalachia, and though companies like Walmart, FedEx, and Coca-Cola are now overhauling their transportation fleets and refrigerators, selling trademarked plant bottles, and putting in energy-efficient light bulbs at their offices to try and "go green," they are still engaged in an unrelenting quest to ship more and more natural resources mined, processed, and packaged in one end of the world to the farthest reaches of the other.

Increasing throughput, the flow of goods through their corporate system, was, at base, what made these companies money. That has been and remains these firms' primary purpose. But that rapid flow of goods through the economy is creating a red-hot planet.

Amazon's Jeff Bezos says he has the answer to this predicament: outer space. In the same 2019 speech where he acknowledged retail and financial infrastructure developed by other firms that made Amazon great, Bezos also claimed that human salvation depended on creating colonies outside Earth's stratosphere. "If we move out into the solar system," Bezos said, "we will have, for all practical purposes, unlimited resources." It was clear from the numbers, Bezos acknowledged, that more efficient technology would not radically reduce our impact on Earth's vital ecological resources if firms kept growing at their current pace. But rather than call for businesses to curb their expansion, Bezos urged those listening to simply move beyond our home planet. "We need to save our Earth," he said, "and we shouldn't give up on a future of dynamism and growth for our grandchildren's grandchildren. We can have both." The key was creating cosmic colonies tethered to a vast interplanetary transportation network that would allow companies to tap unlimited resources. And Bezos was well positioned to profit from this undertaking. After all, he was launching a Blue Origin shuttle project, which he claimed would make space travel affordable for more people in the years to come.[19]

This was the imaginative thinking, the best idea, this billionaire had for addressing the problems our planet now faced—problems his company was complicit in compounding. It seems, in other words, that one of the most powerful business executives running our have-it-now, fly-by-night, and buy-on-credit economy was more invested in an evacuation route from planet Earth than figuring out how to adopt truly transformative business strategies that would allow humans to live comfortably and sustainably in their native habitat.

Which is why, if we hope to see the kind of swift transformations in corporate strategy that are needed to curb some of the worst environmental abuses causing climate change in 2022, government agencies will have to play a much more active role in regulating the big firms that have become the conduits of capitalism in our economy. As we've seen, government helped make firms like Walmart, Coca-Cola, FedEx, Delta, and Bank of America big. In fact, without massive infusions of capital that came from the state, none of these businesses would have had the Earth-changing power they do today. Yet federal laws, from the Clean Air Act to the Clean Water Act, have targeted only the most minor pollution problems these businesses have fueled. Our government can and should demand more of these firms, which don't have the traditional smokestacks and sewer pipelines these older laws were designed to target. Congress could empower the EPA to demand that these businesses harness the sophisticated monitoring technologies they have developed over the years to ensure that toxic compounds or labor abuses do not come packaged with the goods they sell. Likewise, financial regulatory bodies could require firms to be much more transparent in reporting their financing of environmentally destructive industries. Regulators could even go further and ban certain lending practices deemed a threat to the future survival of humanity on the planet. None of this would be easy, especially given the fractious political times in which we live, but given the need to act fast to adjust course on climate change, focusing on firms in these areas of retail, logistics, and finance may have much more effect than targeting individual polluting plants and factories.

But while government agencies must rethink how they regulate big business in the era of Amazon, consumers will have to play a role in this transformation too. Most of us have come to expect that we can get almost anything from anywhere at any time with just the click of a button. And while no one would begrudge a life sciences company from FedExing overnight a temperature-sensitive medicine to a hospital to heal a sick patient, one wonders whether every corner shop on every rural road in the world needs a cold Coca-Cola sitting in a greenhouse-gas-emitting cooler that is on twenty-four hours a day, seven days a week. Robert Woodruff's mantra, "Within arm's reach of desire," has become, in so many ways, the Walmart way, the Amazon way, and, really, our whole way of thinking about consumption. But in the era of climate change, we would do well to recognize that we too are embedded in the history offered here, whether we like it or not. Which is why we will also have to do the hard work of confronting our own expectations of what an Amazon deal really means for the people who make what we consume and the planet on which we all depend.

Acknowledgments

This book would not have been possible without Mark Simpson-Vos, UNC Press's amazing editorial director, who believed in me and this project and encouraged me to write it so many years ago. I first met Mark when I was a young graduate student trying to learn how to become a professional historian. We bonded over basketball and bourbon at the annual conference for the American Society for Environmental History, which typically coincided with March Madness. In the lull between panels and roundtables, Mark and I were often trying to steal away for a moment or two to watch his Tarheels or my Cavaliers advance to the next round in the Big Dance. I knew then that Mark was an amazing mentor and friend, the kind of person who offered up generous amounts of his time to new scholars who knew little about the publishing world. I'm so glad that, many years later, we finally got the chance to work together. The experience was a dream.

UNC Press offered generous funding to complete this project, freeing me from the typical demands of applying for numerous research grants. The Stuart A. Rose Manuscript, Archives, and Rare Book Library at Emory University and the Wilson Special Collections Library at the University of

North Carolina also offered financial support, though in the end the coronavirus pandemic prevented me from taking advantage of these funds. Nevertheless, I relied heavily on records at both archives and am grateful to the staff at these institutions who helped me find critical documents featured herein.

Archivists at the Vermont Historical Society and at the M. E. Grenander Department of Special Collections and Archives at the University at Albany, State University of New York, worked with me to identify key records critical to the Walmart story, while the Delta Flight Museum offered me access to archival resources that were essential for the Delta chapters. The William J. Clinton Digital Presidential Library, the National Archives in College Park, and the Hagley Museum and Library also preserved documents that added richness to the FedEx and Coca-Cola chapters.

I conducted many interviews for this book, and I won't list here all the names of people who dedicated their time to help me get this story right, but a glance through the endnotes will reveal the former chief executives, sustainability officers, company employees, and environmental activists who added their voices to this story. I'm grateful for these interviews, which in so many ways allowed me to understand the inner workings of firms whose archival records often offer only a limited view of the past.

I'm very grateful for the efforts of two referees for UNC Press who offered incredible feedback on early drafts of the manuscript. I'd like to express special appreciation to Edward L. Ayers, Paul Belonick, Stephen Brain, Jerome Buescher, Darren E. Grem, Grace Hale, Mark Hersey, Chris Jones, Catherine McNeur, Chris Reed, Andrew Robichaud, Adam Rome, Drew A. Swanson, and Nathanael Swart, who read whole drafts and/or sections of the book and helped with editing. Finally, two referees for *Environmental History* pushed me to hone and refine arguments that now appear in the banking chapters of this book. I feel very fortunate to have had such tremendous support from colleagues in my field of expertise.

The Department of History and the Sustainability Institute at Ohio State University have been constant backers of my work, and I am so happy to have found an institutional home that has given me the resources and freedom to investigate some of the world's biggest businesses. I am particularly indebted to department chair Scott Levi, who has encouraged me to pursue these big research projects and who has been there for me whenever I needed help.

Finally, I have to thank my family, especially my wife and my boys, for putting up with me over the past several years. I wrote this book at the same time as I was finishing another book on the environmental history of Monsanto. That put an unusual strain on all of us, though I'm happy to look back at all the fond memories we had playing together through the chaos of the pandemic. In so many ways, I labored at this book so hard because, at base, I want a brighter future for the people I love most. Thank you for being my inspiration, Joya, River, and Blue.

Notes

ABBREVIATIONS

AGC Papers	Asa Griggs Candler Papers, Stuart A. Rose Manuscript, Archives, and Rare Book Library, Emory University, Atlanta, Georgia
Delta Archives	Delta Flight Museum, Atlanta, Georgia
Hagley Museum	Cinecraft Productions Films Collection, Audiovisual Collections and Digital Initiatives Department, Hagley Museum and Library, Wilmington, Delaware
McColl Papers	McColl Family Papers, Special Collections, Wilson Library, University of North Carolina, Chapel Hill
NationsBank Records	NationsBank Records, Special Collections, Wilson Library, University of North Carolina, Chapel Hill
OCCA Records	Otsego County Conservation Association Records, M. E. Grenander Department of Special Collections and Archives, University Libraries, University at Albany, State University of New York
RWW Papers	Robert W. Woodruff Papers, Stuart A. Rose Manuscript, Archives, and Rare Book Library, Emory University, Atlanta, Georgia
WCRG Records	Williston Citizens for Responsible Growth Records, 1988–98, Vermont Historical Society, Barre, Vermont

PREFACE

1. Author interview with Richard Smith, regional president of the Americas and executive vice president of FedEx, March 15, 2021.

2. Author interview with Richard Smith.

3. Author interview with Richard Smith.

4. Author interview with Richard Smith; "Country Roads: UPS, FedEx Ramp Up Rural Vaccine Delivery," *FreightWaves*, May 13, 2021, https://www.freightwaves.com/news/country-roads-ups-fedex-ramp-up-rural-vaccine-delivery; FedEx, "FedEx Introduces SenseAware, the Next Generation Supply Chain Information Platform" (press release), November 17, 2009, https://newsroom.fedex.com/newsroom/fedex-introduces-senseaware-the-next-generation-supply-chain-information-platform/; FedEx, "FedEx to Transform Package Tracking with SenseAware ID, the Latest Innovation in FedEx Sensor Technology" (press release), September 14, 2020, https://newsroom.fedex.com/newsroom/senseaware-id/; "FedEx Exec Shows Off 'Guardian Angel' Tracking Technology," CNN Business, December 13, 2020, https://www.cnn.com/videos/business/2020/12/13/fedex-coronavirus-vaccine-tracking-richard-smith-nr-sot-vpx.cnn.

5. FedEx, "FedEx and Microsoft Join Forces to Transform Commerce" (press release), May 18, 2020, https://newsroom.fedex.com/newsroom/fedex-surround/.

6. FedEx, "FedEx Distribution of COVID-19 Vaccines Grows, Reaches 100 Million Vaccine Doses Delivered" (press release), April 1, 2021, https://newsroom.fedex.com/newsroom/global-english/fedex-distribution-of-covid-19-vaccines-grows-reaches-100-million-vaccine-doses-delivered.

7. "Country Roads: UPS, FedEx Ramp Up Rural Vaccine Delivery," *FreightWaves*, May 13, 2021, https://www.freightwaves.com/news/country-roads-ups-fedex-ramp-up-rural-vaccine-delivery; Richard W. Smith, testimony before U.S. Senate Committee on Commerce, Science, and Transportation, Subcommittee on Transportation and Safety, *Hearing on the Logistics of Transporting a COVID-19 Vaccine*, December 10, 2020, https://www.commerce.senate.gov/2020/12/the-logistics-of-transporting-a-covid-19-vaccine.

8. 2022 FedEx ESG Report, 5, 15, 20, 36–37; 2021 FedEx ESG Report, 34.

INTRODUCTION

1. Edward L. Ayers, *Promise of the New South: Life after Reconstruction* (New York: Oxford University Press, 1992), 101–2. For other works that discuss the way Jim Crow segregation was constructed in a time of tremendous economic change, see Grace Elizabeth Hale, *Making Whiteness: The Culture of Segregation in the South, 1890–1940* (New York: Pantheon Books, 1998); C. Vann Woodward, *Origins of the New South, 1877–1913* (Baton Rouge: Louisiana State University Press, 1971).

2. Ayers, *Promise of the New South*, 159.

3. James C. Cobb, *The Selling of the South: The Southern Crusade for Industrial Development, 1936–1990*, 2nd ed. (Urbana: University of Illinois Press, 1993), 1–4, 177, 230, 263–64, 281. On the American South in the Cold War, see Bruce Schulman, *From Cotton Belt to Sunbelt: Federal Policy, Economic Development, and the Transformation of the South, 1938–1980* (New York: Oxford University Press, 1991).

4. Gavin Wright, *Sharing the Prize: The Economics of the Civil Rights Revolution in the American South* (Cambridge, Mass.: Belknap Press of Harvard University Press, 2013), 74, 110.

5. On the ecological history of southern agriculture, see Pete Daniel, *Breaking the Land: The Transformation of Cotton, Tobacco, and Rice Cultures since 1880* (Urbana: University of Illinois Press, 1986); James C. Giesen, *Boll Weevil Blues: Cotton, Myth and Power in the American South* (Chicago: University of Chicago Press, 2011); Mark D. Hersey, *My Work Is That of Conservation: An Environmental Biography of George Washington Carver* (Athens: University of Georgia Press, 2011); Lynn A. Nelson, *Pharsalia: An Environmental Biography of a Southern Plantation* (Athens: University of Georgia Press, 2007); Erin Stewart Mauldin, *Unredeemed Land: An Environmental History of Civil War and Emancipation in the Cotton South* (New York: Oxford University Press, 2018); Hayden R. Smith, *Carolina's Golden Fields: Inland Rice Cultivation in the South Carolina Lowcountry, 1670–1860* (Cambridge: Cambridge University Press, 2019); Mart A. Stewart, *What Nature Suffers to Groe: Life, Labor, and Landscape on the Georgia Coast, 1680–1920* (Athens: University of Georgia Press, 1996); Drew A. Swanson, *A Golden Weed: Tobacco and Environment in the Piedmont South* (New Haven, Conn.: Yale University Press, 2014); Paul Sutter, *Let Us Now Praise Famous Gullies: Providence Canyon and the Soils of the South* (Athens: University of Georgia Press, 2015). On the ecological history of Appalachia, see Drew A. Swanson, *Beyond the Mountains: Commodifying Appalachian Environments* (Athens: University of Georgia Press, 2018); Donald E. Davis, *Where There Are Mountains: An Environmental History of the Southern Appalachians* (Athens: University of Georgia Press, 2000); Chad Montrie, *To Save the Land and People: A History of Opposition to Surface Coal Mining in Appalachia* (Chapel Hill: University of North Carolina Press, 2003); Kathryn Newfont, *Blue Ridge Commons: Environmental Activism and Forest History in Western North Carolina* (Athens: University of Georgia Press, 2012). On timber companies, forest management, and the southern environment, see Albert G. Way, *Conserving Southern Longleaf: Herbert Stoddard and the Rise of Ecological Land Management* (Athens: University of Georgia Press, 2011); and James E. Fickle, *Green Gold: Alabama's Forests and Forest Industries* (Tuscaloosa: University of Alabama Press, 2014). On the environmental history of peach farming, see William Thomas Okie, *The Georgia Peach: Culture, Agriculture, and Environment in the American South* (Cambridge: Cambridge University Press, 2016). On rivers and wetlands, see Christopher Manganiello, *Southern Water, Southern Power: How the Politics of Cheap Energy and Water Scarcity Shaped a Region* (Chapel Hill: University of North Carolina Press, 2015); Chris Morris, *The Big Muddy: An Environmental History of the Mississippi and Its Peoples from Hernando de Soto to Hurricane Katrina* (New York: Oxford University Press, 2012); Jack E. Davis, *The Gulf: The Making of an American Sea* (New York: Liveright, 2017); Jack E. Davis and Raymond Arsenault, *Paradise Lost? The Environmental History of Florida* (Gainesville: University Press of Florida, 2005); Ari Kelman, *A River and Its City: The Nature of Landscape in New Orleans* (Berkeley: University of California Press, 2006); and Adam Mandelman, *The Place with No Edge: An Intimate History of People, Technology, and the Mississippi River Delta* (Baton Rouge: Louisiana State University Press, 2020). William D. Bryan has explored the ways in which New South businessmen promoted the conservation of southern natural resources in a quest for "permanence" and perpetual economic growth. See William D. Bryan, *The Price of Permanence: Nature and*

Business in the New South (Athens: University of Georgia Press, 2018). Race and labor are front and center in many southern environmental histories, especially in works such as Monica R. Gisolfi's investigation of the human health and environmental costs of chicken farming in Georgia and Ellen Griffith Spears's brilliant work on Black and white working-class citizens who were harmed by a Monsanto chemical plant in Anniston, Alabama. Dianne D. Glave has produced important works that reclaim stories of Black environmentalists in the American South and beyond. See Monica R. Gisolfi, *The Takeover: Chicken Farming and the Roots of American Agribusiness* (Athens: University of Georgia Press, 2017); Ellen Griffith Spears, *Baptized in PCBs: Race, Pollution, and Justice in an All-American Town* (Chapel Hill: University of North Carolina Press, 2014); Dianne D. Glave, *Rooted in the Earth: Reclaiming the African American Environmental Heritage* (Chicago: Lawrence Hill Books, 2010); Dianne D. Glave and Mark Stoll, eds., *"To Love the Wind and the Rain": African Americans and Environmental History* (Pittsburgh, Pa.: University of Pittsburgh Press, 2006). On the environmental history of the American Civil War, see Lisa M. Brady, *War upon the Land: Military Strategy and the Transformation of Southern Landscapes during the American Civil War* (Athens: University of Georgia Press, 2012); Adam Wesley Dean, *An Agrarian Republic: Farming, Antislavery Politics, and Nature Parks in the Civil War Era* (Chapel Hill: University of North Carolina Press, 2015); Brian Allen Drake, ed., *The Blue, the Gray, and the Green: Toward an Environmental History of the Civil War* (Athens: University of Georgia Press, 2014); Kathryn Shively Meier, *Nature's Civil War: Common Soldiers and the Environment in 1862 Virginia* (Chapel Hill: University of North Carolina Press, 2013); and Judkin Browning and Timothy Silver, *An Environmental History of the Civil War* (Chapel Hill: University of North Carolina Press, 2020). For sweeping surveys of southern environmental history, see Timothy H. Silver, *A New Face on the Countryside: Indians, Colonists, and Slaves in South Atlantic Forests, 1500–1800* (Cambridge: Cambridge University Press, 1990); Alfred E. Cowdrey, *This Land, This South: An Environmental History* (Lexington: University of Kentucky Press, 1996); Jack Temple Kirby, *Mockingbird Song: Ecological Landscapes of the South* (Chapel Hill: University of North Carolina Press, 2006); and Donald E. Davis, Craig E. Colten, Megan Kate Nelson, Barbara L. Allen, and Mikko Saikku, eds., *Southern United States: An Environmental History* (Santa Barbara: ABC-CLIO, 2006).

6. In his review of scholarship on the American South and the world, Tore C. Olsson ("The South in the World since 1865: A Review Essay," *Journal of Southern History* 87, no. 1 [February 2021]: 67–108) mentions several works that highlight the power of southern firms to shape global markets, including Nan Enstad, *Cigarettes, Inc.: An Intimate History of Corporate Imperialism* (Chicago: University of Chicago Press, 2018); B. Alex Beasley, *Expert Capital: Houston and the Making of a Service Empire* (Cambridge, Mass.: Harvard University Press, forthcoming); Bethany Moreton, *To Serve God and Wal-Mart: The Making of Christian Free Enterprise* (Cambridge, Mass.: Harvard University Press, 2009); and Wanda Rushing, *Memphis and the Politics of Place: Globalization in the American South* (Chapel Hill: University of North Carolina Press, 2009). *Country Capitalism* seeks to add to this burgeoning literature, focusing on the global ecological impacts of southern multinational firms. Though *Country Capitalism* is focused on the twentieth century, it is worth noting that a robust literature has also emerged over the past several decades on the history of American slavery that has shown how integral southern cotton plantations were to global capital markets and the making of American

capitalism. For example, see Edward E. Baptist, *The Half Has Never Been Told: Slavery and the Making of American Capitalism* (New York: Basic Books, 2014); Sven Beckert, *Empire of Cotton: A New History of Global Capitalism* (London: Penguin Books, 2015); Joshua D. Rothman, *Flush Times and Fever Dreams: A Story of Capitalism and Slavery in the Age of Jackson* (Athens: University of Georgia Press, 2012); and Jack Lawrence Schermerhorn, *The Business of Slavery and the Rise of American Capitalism, 1815–1860* (New Haven, Conn.: Yale University Press, 2015).

7. I would like to thank environmental historian Stephen J. Macekura for suggesting the term "commercial ecology."

8. On the history of Chicago and the link between countryside and city economies, see William Cronon, *Nature's Metropolis: Chicago and the Great West* (New York: W. W. Norton, 1991).

9. James C. Cobb, *The South and America since World War II* (New York: Oxford University Press, 2011), 54.

10. See "History: Urban and Rural Areas," U.S. Census Bureau, accessed October 25, 2022, https://www.census.gov/history/.

11. On the complex migration patterns that shaped the American South from 1790 to 2020, see Edward L. Ayers, *Southern Journey: The Migrations of the American South, 1790–2020* (Baton Rouge: Louisiana State University Press, 2020).

12. As southern historian Andrew C. Baker has noted, "Finding the pristine countryside before metropolitan engagement is as impossible as finding pristine nature." Andrew C. Baker, *Bulldozer Revolutions: A Rural History of the Metropolitan South* (Athens: University of Georgia Press, 2018), 4.

13. "Rural America," U.S. Census Bureau, accessed November 3, 2022, https://mtgis-portal.geo.census.gov/arcgis/apps/MapSeries/index.html?appid=49cd4bc9c8eb444ab51218c1d5001ef6.

14. Andrew C. Baker offers another nice framing when thinking about a definition for the "countryside," noting that both rural businessmen and metropolitan entrepreneurs saw the countryside not as a distinct place but as a "marketable product." Baker, *Bulldozer Revolutions*, 5.

15. Journalist Brad Stone revealed that Bezos read and admired Sam Walton's autobiography, and Stone also explained in detail how Amazon hired Walmart employees to help the Seattle firm develop its distribution network. See Brad Stone, *The Everything Store: Jeff Bezos and the Age of Amazon* (New York: Little, Brown, 2013), 60–62, 72–75, 116–19, 125, 131–32, 161–62, 171, 245–46. For a discussion of the differences and similarities between Amazon's and Walmart's business strategies, see Colby Ronald Chiles and Marguarette Thi Dau, "An Analysis of Current Supply Chain Best Practices in the Retail Industry with Case Studies of Wal-Mart and Amazon.com" (master's thesis, Massachusetts Institute of Technology, June 2005), 169, 172–75.

16. Nicolás Rivero, "Amazon's Ambitions to Build an Air Freight Empire Got a Lift from the Pandemic," *Quartz*, December 22, 2021, https://qz.com/.

17. Jeff Bezos, "The Purpose of Going to Space," speech, Blue Origin event, Washington, D.C., May 9, 2019, in *Invent and Wander: The Collected Writings of Jeff Bezos* (Cambridge, Mass.: Harvard Business Review Press, 2020), 250.

18. Melody Brue, "Is Amazon Building the Next Generation Bank," *Forbes*, April 30, 2021, www.forbes.com/.

CHAPTER 1

1. Kathryn W. Kemp, *God's Capitalist: Asa Candler of Coca-Cola* (Macon, Ga.: Mercer University Press, 2002), 1–18; Mark Pendergrast, *For God, Country and Coca-Cola: The Definitive History of the Great American Soft Drink and the Company That Makes It*, 3rd ed. (New York: Basic Books, 2013), 45–46.

2. "Asa Griggs Candler: Founder of the Coca-Cola Company," *Coca-Cola Bottler*, April 1959, 55.

3. Pendergrast, *For God, Country and Coca-Cola*, 45–46; Kemp, *God's Capitalist*, 11–12.

4. Letter from Asa G. Candler to Asa W. Griggs, September 11, 1872, box 1, folder 1: Correspondence: 1872, Sept. 11–1897, Dec. 2, AGC Papers; Kemp, *God's Capitalist*, 14–16.

5. Letter from Asa G. Candler to Asa W. Griggs, September 11, 1872, box 1, folder 1: Correspondence: 1872, Sept. 11–1897, Dec. 2, AGC Papers.

6. U.S. Department of the Interior, Census Office, *Ninth Census*, vol. 1, *The Statistics of the Population of the United States [. . .]* (Washington, D.C.: Government Printing Office, 1872), 99–100.

7. Yolande Gwin, "Home in Inman Park Was Asa Candler's Favorite," *Atlanta Journal and Constitution*, March 12, 1972, 14-G; Frederick Allen, *Secret Formula: How Brilliant Marketing and Relentless Salesmanship Made Coca-Cola the Best-Known Product in the World* (New York: HarperBusiness, 1994), 32; Pendergrast, *For God, Country and Coca-Cola*, 47.

8. "Croup!," *Atlanta Constitution*, September 13, 1885, 9.

9. Pendergrast, *For God, Country and Coca-Cola*, 49; "Asa Griggs Candler: Founder of the Coca-Cola Company," *Coca-Cola Bottler*, April 1959, 55; letter from Asa G. Candler to family, May 27, 1885, box 1, folder 1, AGC Papers; "Holmes' Sure Cure Mouth Wash and Dentifrice," *Selma (Ala.) Times*, November 1, 1884, 3; "Blood Balm: Atlanta Druggists," *Daily American* (Nashville, Tenn.), December 24, 1884, 6; "Cures Blood Poison," *Junction City (Kans.) Republican*, December 14, 1882, 4.

10. Letter from Asa Candler, 1880s [exact date unclear], box 1, folder 1, AGC Papers.

11. Pendergrast, *For God, Country and Coca-Cola*, 18–22; Bartow Elmore, *Citizen Coke: The Making of Coca-Cola Capitalism* (New York: W. W. Norton, 2015), 18–19; Census Office, *Ninth Census*, 1:105.

12. Elmore, *Citizen Coke*, 20; "Who Is Mariani? How He Came from Corsica and Grew to Be One of the Great Men of Paris," *Washington Post*, May 31, 1898, 7; Allen, *Secret Formula*, 23; Pendergrast, *For God, Country and Coca-Cola*, 22–23.

13. Pendergrast, *For God, Country and Coca-Cola*, 17–23; Elmore, *Citizen Coke*, 18–21.

14. "The Best Nerve Tonic," *Atlanta Constitution*, June 21, 1885, 7; Pendergrast, *For God, Country and Coca-Cola*, 22, 24. On the history of kola nut cultivation in West Africa, see Edmund Abaka, *Kola Is God's Gift: Agricultural Production, Export Initiatives, and the Kola Industry of Asante and the Gold Coast c. 1820–1950* (Athens: Ohio University Press, 2005); and Paul E. Lovejoy, "Kola in the History of West Africa (La kola dans l'histoire de l'Afrique occidentale)," *Cahiers d'études Africaines* 20 (1980): 97–134.

15. "Coca-Cola, the Ideal Brain Tonic," *Montgomery Daily Advertiser*, April 17, 1887, 5; Elmore, *Citizen Coke*, 21–22; Pendergrast, *For God, Country and Coca-Cola*, 24, 26–27. For a different take on why Pemberton created his temperance drink that suggests the lower price point was a major factor, see Allen, *Secret Formula*, 26.

16. "Success of the Wonderful Headache Specific, Coca-Cola," *Memphis Daily Appeal*, June 29, 1887, 6; letter from Asa Candler, 1880s [exact date unclear], box 1, folder 1, AGC Papers.

17. "Asa Griggs Candler: Founder of the Coca-Cola Company," *Coca-Cola Bottler*, April 1959, 185; Pendergrast, *For God, Country and Coca-Cola*, 32–44; Elmore, *Citizen Coke*, 24–25.

18. Pendergrast, *For God, Country and Coca-Cola*, 50–52; 1922 Coca-Cola Annual Report, 14; Richard Tedlow, *New and Improved: The Story of Mass Marketing in America* (Boston: Harvard Business School Press, 1996), 36. Asa Candler talked about "cotton buyers" that had "traveled for us two seasons" in a letter to Howard Candler, March 13, 1899, box 1, folder 2, AGC Papers.

19. Wilbur G. Kurtz Jr., "Joseph A. Biedenharn," *Coca-Cola Bottler*, August 1944, 15–18; Franklin M. Garrett, "Coca-Cola in Bottles," *Coca-Cola Bottler*, April 1959, 79–80; "Joseph A. Biedenharn, Bottling Pioneer, Dies at 85," *Coca-Cola Bottler*, November 1952, 25.

20. Kurtz, "Joseph A. Biedenharn," 18.

21. "Founder of the Coca-Cola Company," 185–87; Kurtz, "Joseph A. Biedenharn," 15–18; letter from Asa Candler to Howard Candler, 1899, box 1, folder 2, AGC Papers; 1922 Coca-Cola Annual Report, 14.

22. "Valdosta Has Bottled Coca-Cola since 1897," *Coca-Cola Bottler*, July 1955, 41; Garrett, "Coca-Cola in Bottles," 80; Franklin M. Garrett, "Benjamin Franklin Thomas," *Coca-Cola Bottler*, April 1959, 85–86; "Joseph Brown Whitehead," *Coca-Cola Bottler*, April 1959, 87–89; Elmore, *Citizen Coke*, 28–32; Pendergrast, *For God, Country and Coca-Cola*, 66–70.

23. Garrett, "Coca-Cola in Bottles," 79–84; "Clarksville, Tennessee," *Coca-Cola Bottler*, January 1956, 22; "Greenwood, Mississippi," *Coca-Cola Bottler*, January 1956, 25; Roy S. Jones, "A Brief History of Coca-Cola Overseas," *Coca-Cola Bottler*, April 1959, 181; J. J. Willard, "Some Early History of Coca-Cola Bottling," *Coca-Cola Bottler*, August 1944, 27; U.S. Census Bureau, *Thirteenth Census of the United States*, vol. 2, *Population, Reports by States . . . Alabama–Montana*, 1035, and vol. 2, *Population, Reports by States . . . Nebraska–Wyoming*, 735, www.census.gov.

24. Garrett, "Coca-Cola in Bottles," 81.

25. Michael M. Cohen, "Jim Crow's Drug War: Race, Coca-Cola, and the Southern Origins of Drug Prohibition," *Southern Cultures* 12, no. 3 (Fall 2006): 70; Elmore, *Citizen Coke*, 115–16.

26. See, for example, the shift in the company letterhead: in April 1905, it included references to the "wonderful coca plant and famous cola nut," whereas in May 1905 and after it simply featured company offices in big cities. Letter from Asa G. Candler to Sissie, April 12, 1905; letter from Asa G. Candler to Warren Candler, May 15, 1905, box 1, folder: Correspondence: 1902, Feb. 1–1907, Aug. 5, AGC Papers.

27. Letter from Asa Candler to Howard Candler, July 27, 1908, box 1, folder: Correspondence: 1908, June 16–1912, Dec. 6, AGC Papers.

28. For an example of Candler's letterhead in 1895, see letter from Asa Candler to Howard Candler, March 7, 1895, box 1, folder 1, AGC Papers. References to coca and kola nuts seem to disappear around the spring of 1905. See letter from Asa Candler to Warren Candler, May 15, 1905, box 1, folder: Correspondence: 1902, Feb. 1–1907, Aug. 5, ACG Papers.

29. Letter from Asa Candler to Howard Candler, June 2, 1902; letter from Asa

Candler to Howard Candler, May 29, 1902; box 1, folder: Correspondence: 1902, Feb. 1–1907, Aug. 5, AGC Papers.

30. Asa Candler, keynote speech, Atlanta National Automobile Exposition, November 6, 1909, box 2, folder 2, AGC Papers.

31. Asa Candler, keynote speech.

32. Asa Candler, keynote speech; "Knoxville Celebrates Its 50th Anniversary," *Coca-Cola Bottler*, November 1952, 29; "Asheville Now in 41st Year," *Coca-Cola Bottler*, March 1946, 23; "Palestine, Texas," *Coca-Cola Bottler*, January 1956, 41; "Columbus, Ohio," *Coca-Cola Bottler*, January 1956, 23; "Another Louisville Success That Always Gets There," *Courier Journal* (Louisville, Ky.), June 14, 1921, 3. On the poor condition of southern roadways before 1915, see Tammy Ingram, *Dixie Highway: Road Building and the Making of the Modern South, 1900–1930* (Chapel Hill: University of North Carolina Press, 2014), 13–42.

33. Letter from Howard Candler to Asa Candler, March 1, 1909, box 1, folder: Correspondence: 1908, June 16–1912, Dec. 6, AGC Papers; "Founder of the Coca-Cola Company," 188; 1922 Coca-Cola Annual Report, 14.

34. Elmore, *Citizen Coke*, 77; Allen, *Secret Formula*, 104. On the ecological consequences of Havemeyer's operations in Cuba, see Richard Tucker, *Insatiable Appetite: The United States and the Ecological Degradation of the Tropical World* (Berkeley: University of California Press, 2000), 37–48.

35. Elmore, *Citizen Coke*, 94–95.

36. Asa Candler, untitled speech, n.d., box 2, folder 29, AGC Papers; Asa Candler, "Southern Patriotism in Business Endeavor," speech, n.d., box 2, folder 30, AGC Papers.

37. Asa Candler, untitled speech to the Atlanta Chamber of Commerce, n.d., box 2, folder 1, AGC Papers; Asa Candler, untitled speech, n.d., box 2, folder 5, AGC Papers; Candler, "Southern Patriotism in Business Endeavor."

38. Asa Candler, untitled speech, n.d., box 2, folder 29, AGC Papers; Candler, untitled speech to the Atlanta Chamber of Commerce, n.d., box 2, folder 1, AGC Papers; Asa Candler, "The South's Commerce—Past and Future," speech, n.d., box 2, folder 33, AGC Papers.

39. Historian Nelson Lichtenstein notes that the term "logistics" mainly applied to military operations prior to the 1980s. See Nelson Lichtenstein, "Walmart's Long March to China: How a Mid-American Retailer Came to Stake Its Future on the Chinese Economy," in *Walmart in China*, ed. Anita Chan (Ithaca, N.Y.: Cornell University Press, 2011), 22.

40. Document detailing Coca-Cola's financial history from 1886 to 1919, series 3, box 371, folder 13, RWW Papers; Allen, *Secret Formula*, 64.

41. Letter from Asa Candler to Warren Candler, January 11, 1913; letter from Asa Candler to Warren Candler, June 29, 1913, box 1, folder: Correspondence: 1913, Jan. 1–1914, July 10, AGC Papers; Pendergrast, *For God, Country and Coca-Cola*, 118.

42. Frederick Allen, "Robert Woodruff: His Power Is Subtle, His Power Is Coke," *Atlanta Journal and Constitution*, April 16, 1977, 4-A; document detailing Coca-Cola's financial history from 1886 to 1919, series 3, box 371, folder 13, RWW Papers.

43. Elmore, *Citizen Coke*, 89–90; Pendergrast, *For God, Country and Coca-Cola*, 146.

44. Charles Elliot, *Robert Winship Woodruff: A Biography of the "Boss"* (Robert W. Woodruff estate, 1979), 19–20, 81, 86–87.

45. Associated Press, "Robert Woodruff Is Dead: Created Coca-Cola Empire," *New York Times*, March 9, 1985, 16; Elliot, *Robert Winship Woodruff*, 22; Pendergrast, *For*

God, Country and Coca-Cola, 145. It's worth noting that in the 1960s White Motor Company became directly involved in a kind of country capitalism of another sort: servicing the countryside by developing farm equipment for agribusiness. This division of the firm later became part of Allis-Gleaner Corporation (AGCO), which is headquartered in Duluth, Georgia. In 2012, AGCO was the third-biggest farm equipment business on the planet. See Lisa M. Fine, "The 'Fall' of Reo in Lansing, Michigan, 1955–1975," in *Beyond Ruins: The Meanings of Deindustrialization*, ed. Jefferson Cowie and Joseph Heathcott (Ithaca, N.Y.: ILR Press, 2003), 50; William A. Dando, *Food and Famine in the 21st Century* (Santa Barbara, Calif.: ABC-CLIO, 2012), 421; Robert N. Pripps, *The Field Guide to Classic Farm Tractors* (Beverly, Mass.: Motorbooks, 2020), 227.

46. "Robert Woodruff Wins Great Honor from the White Co.," *Atlanta Constitution*, September 2, 1917, B7; "Two Atlanta Men Honored by White Motor Company," *Atlanta Constitution*, October 4, 1929, 5; Elliot, *Robert Winship Woodruff*, 93.

47. Elliot, *Robert Winship Woodruff*, 96; "Robert W. Woodruff Promoted to Major," *Atlanta Constitution*, November 1, 1918, 9; Tom Seegmueller, "Coke Mastermind Woodruff Had Huge Impact on Southwest Georgia," *Albany (Ga.) Herald*, February 1, 2020, www.albanyherald.com/features/.

48. "Roll Call of White Truck Fleet in Actual Service," *Philadelphia Inquirer*, May 4, 1921, 15; Pendergrast, *For God, Country and Coca-Cola*, 154; Arthur P. Pratt, "My Life with Coca-Cola," *Coca-Cola Bottler*, April 1959, 176; Allen, *Secret Formula*, 154.

49. Ingram, *Dixie Highway*, 11.

50. Allen, *Secret Formula*, 130–31, 150–54; Pendergrast, *For God, Country and Coca-Cola*, 146.

51. Pendergrast, *For God, Country and Coca-Cola*, 154; Elmore, *Citizen Coke*, 101.

52. 1929 Coca-Cola Annual Report, 14; Pendergrast, *For God, Country and Coca-Cola*, 153. "After National Distribution—What?" is from a 1929 article penned by Robert Woodruff and quoted by Pendergrast on page 153 of *For God, Country and Coca-Cola*.

53. Pendergrast, *For God, Country and Coca-Cola*, 184–86; Elmore, *Citizen Coke*, 158.

54. Elmore, *Citizen Coke*, 160; Hays, *Real Thing*, 81–82.

55. Edward L. Ayers, *Southern Journey: The Migrations of the American South, 1790–2020* (Baton Rouge: Louisiana State University Press, 2020), 84–87; James C. Cobb, *The South and America since World War II* (New York: Oxford University Press, 2011), 54.

56. Ayers, *Southern Journey*, 84–87; Cobb, *The South and America*, 54.

57. Letter from Major Dean N. Walker, Medical Corps, Office of the Base Surgeon, Headquarters Trinidad Base Command, Trinidad, British West Indies, to Aubrey C. Boyce, Canning & Company, Ltd., September 29, 1941; letter from Major Rupert F. Anderson to Coca-Cola Company, Atlanta, Georgia, n.d.; letter from Major General Ralph Pennell, Headquarters Hawaii Division, Office of the Representative of the Military Governor of the Territory of Hawaii, to War Production Board, San Francisco, Calif., May 28, 1942; letter from Lieutenant Colonel John F. Neu, War Department, Headquarters Force "A," Office of the Quartermaster, Paramaribo, Dutch Guiana, box 588, entry 59, Rationing Department, National Office, Food Rationing Division, Office of the Director, General Correspondence, 1942–45, Cla-Coc, Record Group 188: Records of the Office of Price Administration, National Archives and Records Administration, College Park, Md.

58. This is the central thesis of Elmore, *Citizen Coke*.

59. "The Overseas Story," *Coca-Cola Overseas*, June 1948, 5; "The Sun Never Sets on Cacoola," *Time*, May 15, 1950, http://content.time.com.

60. "Sun Never Sets on Cacoola."

CHAPTER 2

1. Author interview with Jeff Seabright, April 29, 2021; Coca-Cola, "Coca-Cola Commits to Climate-Friendly Refrigeration through Engagement with Greenpeace" (press release), December 3, 2009, https://investors.coca-colacompany.com/news-events/press-releases.

2. Author interview with Jeff Seabright. For more on the Kerala crisis, see Bartow Elmore, *Citizen Coke: The Making of Coca-Cola Capitalism* (New York: W. W. Norton, 2015), 155–56, 188.

3. Author interview with Jeff Seabright; Coca-Cola, "Coca-Cola Commits"; "Global Warming Potentials (IPCC Second Assessment Report)," United Nations Climate Change, accessed October 23, 2022, https://unfccc.int/process/transparency-and-reporting/greenhouse-gas-data/greenhouse-gas-data-unfccc/global-warming-potentials.

4. Author interview with Jeff Seabright.

5. Author interview with Jeff Seabright.

6. Author interview with Jeff Seabright.

7. Author interview with Jeff Seabright.

8. 2007/2008 Coca-Cola Sustainability Review Report, 44.

9. "The Morning Reporter," *Daily Reporter* (Independence, Kans.), November 18, 1897, 2; Mark Pendergrast, *For God, Country and Coca-Cola: The Definitive History of the Great American Soft Drink and the Company That Makes It*, 3rd ed. (New York: Basic Books, 2013), 138. On the links between hot temperatures and Coca-Cola consumption, see Lisa Oberlander and Ximena Játiva, "How Heat Waves Increase Your Cravings for Sodas: Findings from Mexico," *The Conversation*, July 12, 2019, https://theconversation.com/how-heat-waves-increase-your-craving-for-sodas-findings-from-mexico-119351.

10. "New Plants and Improvements," *Ice and Refrigeration* 34 (February 1908): 80; "News of Companies and Plants," *Cold Storage and Ice Trade* 43 (April 1912): 88; "Coca-Cola Bottler Tells of Refrigeration Value," *Altoona (Pa.) Tribune*, August 12, 1931, 11.

11. Author interview with Coca-Cola employee at Coca-Cola Bottling Company United, headquartered in Birmingham, Ala., April 2, 2021; "Coca-Cola to Hold Open House," *Clarksville (Tenn.) Leaf-Chronicle*, October 17, 1938, 7.

12. Jonathan Rees, *Refrigeration Nation: A History of Ice, Appliances, and Enterprise in America* (Baltimore: Johns Hopkins University Press, 2013), 44–51, 72.

13. "New Plants and Improvements," *Ice and Refrigeration* 48 (April 1915): 272; "News of Companies and Plants," 88; Rees, *Refrigeration Nation*, 159; Björn Palm, "Hydrocarbons as Refrigerants in Small Heat Pump and Refrigeration Systems—a Review," *International Journal of Refrigeration* 31 (2008): 553; "Coca-Cola Gets $1,200 Refrigeration Permit," *Wilkes-Barre (Pa.) Record*, July 20, 1936, 3; "Coca-Cola to Hold Open House."

14. Rees, *Refrigeration Nation*, 178, 148, 164–65.

15. Rees, *Refrigeration Nation*, 148, 165–66.

16. Rees, *Refrigeration Nation*, 160–61.

17. Constance L. Hays, *The Real Thing: Truth and Power at the Coca-Cola Company* (New York: Random House, 2004), 11; 1929 Coca-Cola Annual Report, 14; Carroll Gantz, *Refrigeration: A History* (Jefferson, N.C.: McFarland and Company, 2015), 150; Kerry Segrave, *Vending Machines: An American Social History* (Jefferson, N.C.: McFarland and Company, 2002), 63; Westinghouse Electric & Manufacturing Company and Cinecraft, Inc., Coolers for Coca-Cola production script, box 19, folder 1, Hagley Museum.

18. Westinghouse Electric & Manufacturing Company, training program for Coca-Cola salesman script, 1943, box 18, folder 17, Hagley Museum; Segrave, *Vending Machines*, 63.

19. On Coca-Cola advertisement's featuring Woodruff's "arm's reach of desire" mantra, see, for example, "This Morning, Tomorrow Morning and Every Morning," *Daily News* (New York), June 28, 1941, 7.

20. "The Overseas Story," *Coca-Cola Overseas*, June 1948, 5.

21. Amy Larkin, *Environmental Debt: The Hidden Costs of the Changing Global Economy* (New York: Palgrave Macmillan, 2013), 32, 107; Mario J. Molina and F. S. Rowland, "Stratospheric Sink for Chlorofluoromethanes: Chlorine Atom-Catalysed Destruction of Ozone," *Nature* 249 (1974): 810–12.

22. Björn Palm, "Hydrocarbons as Refrigerants in Small Heat Pump and Refrigeration Systems—a Review," *International Journal of Refrigeration* 31 (2008): 553; Larkin, *Environmental Debt*, 108.

23. Osmann Sari and Mohamed Balli, "From Conventional to Magnetic Refrigerator Technology," *International Journal of Refrigeration* 37 (2014): 10; author interview with Wolo Lohbeck, Greenpeace activist, January 27, 2021. Lohbeck was a central figure who helped negotiate deals between Greenpeace and major refrigerator manufacturers to help popularize Greenfreeze technology in Europe. Larkin, *Environmental Debt*, 108–10; "Global Warming Potentials."

24. Author interview with Wolo Lohbeck; author interview with Amy Larkin, former Greenpeace strategist, January 18, 2021; John Maté, "Making a Difference: A Case Study of the Greenpeace Ozone Campaign," *RECEIL* 10, no. 2 (2001): 196; Edwin R. Stafford, Cathy L. Hartman, and Yin Liang, "Forces Driving Environmental Innovation Diffusion in China: The Case of Greenfreeze," *Business Horizons* 46, no. 2 (March–April 2003): 49.

25. 1991 Coca-Cola Annual Report, 26–27; author interview with Bryan Jacobs, former director of climate protection at Coca-Cola, January 27, 2021.

26. Author interview with Bryan Jacobs.

27. Author interview with Bryan Jacobs.

28. Author interview with Bryan Jacobs.

29. Author interview with Bryan Jacobs.

30. Maté, "Making a Difference," 196–98; Saad Dishad, Ali Raza Kalair, and Nasrullah Khan, "Review of Carbon Dioxide (CO2) Based Heating and Cooling Technologies: Past, Present, and Future Outlook," *International Journal of Energy Research* 44 (2020): 1413; Stafford, Hartman, and Liang, "Forces Driving Environmental Innovation," 10; author interview with Wolo Lohbeck.

31. "Coca-Cola Strives to Rival Tap Water," *Financial Post* (Toronto, Canada), October 30, 1997, 77; 1996 Coca-Cola Annual Report, 23.

32. Author interview with Corin Mallais, former Greenpeace campaigner, February 25, 2021; author interview with John Maté, former Greenpeace campaigner, January 27, 2021; author interview with Bryan Jacobs; "Activists Try to Shake Up Coke Over Olympic Refrigeration," *GreenBiz*, June 7, 2000, https://www.greenbiz.com/article/activists-try-shake-coke-over-olympic-refrigeration. On the history of Greenpeace, see Frank Zelko, *Make It a Green Peace! The Rise of Countercultural Environmentalism* (New York: Oxford University Press, 2013).

33. Author interview with Corin Mallais; author interview with John Maté; Maté, "Making a Difference," 197; United Nations Environment Programme, "Greenpeace Victory: Coca-Cola Cleans Up," *OzoNews*, UNEP DTIE OzonAction Programme, June 30, 2000, https://wedocs.unep.org/handle/20.500.11822/29941.

34. Author interview with Bryan Jacobs; Blair Palese et al., *How Green the Games? Greenpeace's Environmental Assessment of the Sydney 2000 Olympics*, pamphlet, Greenpeace International and Greenpeace Australia Pacific, September 2000, 32, www.outdoorinov8.com/design/How%20green%20the%20Games.pdf.

35. Author interview with Bryan Jacobs.

36. Isabelle Gerretsen, "How Your Fridge Is Heating Up the Planet," BBC, December 7, 2020, www.bbc.com/future/; author interview with Jeff Seabright.

37. Author interview with Bryan Jacobs.

38. Coca-Cola, "Coca-Cola Installs 1 Millionth HFC-Free Cooler Globally, Preventing 5.25 MM Metric Tons of CO2" (press release), Business Wire, January 22, 2014, www.businesswire.com/news/; author interview with Amy Larkin.

39. Author interview with Amy Larkin; author interview with Jeff Seabright; author interview with Bryan Jacobs.

40. Author interview with Amy Larkin; author email correspondence with Enesta Jones, U.S. Environmental Protection Agency, May 10, 2021.

41. Author interview with Jeff Seabright.

42. Author interview with Jeff Seabright.

43. Author interview with Amy Larkin.

44. Coca-Cola, "Coca-Cola Installs"; Refrigerants Naturally!, "Refrigerants, Naturally! Calls on Parties to Seize the Opportunity Presented by the HFC Amendment for an Accelerated Transition to Natural Refrigerants" (press release), October 2016, www.refrigerantsnaturally.com/.

45. White House, "Fact Sheet: Obama Administration Partners with Private Sector on New Commitments to Slash Emissions of Potent Greenhouse Gases and Catalyze Global HFC Phase Down" (press release), September 16, 2014; 2019 Coca-Cola Business and Sustainability Report, 65.

46. Michael Garry, "Hydrocarbons for Coke," *Accelerate America*, March 2017, 41, https://issuu.com/shecco/docs/aa1703/41.

47. 2018 Coca-Cola Carbon Disclosure Project (CDP) Report, 19, www.coca-colacompany.com/content/dam/journey/us/en/policies/pdf/sustainability/2018-cdp-climate-change-response.pdf; 2018 Coca-Cola Business and Sustainability Report, 44.

48. Author interview with Jeff Seabright. For more on IMAGINE, see "The Story," IMAGINE, accessed November 1, 2022, https://imagine.one/about.

49. Author interview with Jeff Seabright.

50. "Joseph A. Biedenharn, Bottling Pioneer, Dies at 85," *Coca-Cola Bottler*, November 1952, 25.

CHAPTER 3

1. Katia Hetter and Karla Cripps, "The World's Busiest Airport Is . . .," CNN, September 9, 2016, www.cnn.com/2016/09/09/aviation/worlds-busiest-airports-final-2015/; Hartsfield-Jackson Atlanta International Airport website, www.atl.com/about-atl/.

2. Louis Williams, "William Berry Hartsfield and Atlanta Politics: The Formative Years of an Urban Reformer," *Georgia Historical Quarterly* 84, no. 4 (Winter 2000): 656; Rebecca Burns, "Editor's Note," *Atlanta Magazine*, April 2007, 14; Kathryn W. Kemp, *God's Capitalist: Asa Candler of Coca-Cola* (Macon, Ga.: Mercer University Press, 2002), 96–97; Anita Price Davis, *The Margaret Mitchell Encyclopedia* (Jefferson, N.C.: McFarland & Company, 2013), 110.

3. "Timeline: Delta's Trademarks and Slogans," Delta Air Lines, January 1, 2016, https://news.delta.com/timeline-deltas-trademarks-and-slogans.

4. 2015 Delta Air Lines Corporate Responsibility Report, 4–12; 2012 Delta Air Lines 10-K filing, 4; 2018 Delta Air Lines Corporate Responsibility Report, 11; Kate Kelly, "Delta Eyeing ConocoPhillips Refinery for More than $100 Million," CNBC, April 4, 2012, www.cnbc.com/.

5. United Nations Intergovernmental Panel on Climate Change, *Climate Change 2022: Mitigation of Climate Change*, April 2022, TS-67, 10–58, https://www.ipcc.ch/report/ar6/wg3/; David McCollum, Gregory Gould, and David Greene, *Greenhouse Gas Emissions from Aviation and Marine Transportation: Mitigation Potential and Policies*, December 2009, Solutions White Papers Series, Pew Center on Global Climate Change, 5. See also Joyce E. Penner et al., eds., *Summary for Policymakers: Aviation and the Global Atmosphere: A Special Report of IPCC Working Groups I and III* (Cambridge: Cambridge University Press, 1999).

6. U.S. Department of Transportation, Bureau of Transportation Statistics, *Freight Shipments in America: Preliminary Highlights from the 2002 Commodity Flow Survey* (Washington, D.C.: Bureau of Transportation Statistics, 2002), 7–9, 19–22; "IAG Backs Net Carbon Emissions by 2050," *FreightWaves*, October 14, 2019, https://www.freightwaves.com/news/iag-backs-net-zero-carbon-emissions-by-2050.

7. "Top 23 Largest Airlines Company Rankings by Total Assets," Sovereign Wealth Fund Institute, accessed October 14, 2022, https://www.swfinstitute.org/fund-rankings/airlines-company; "Airline 50 2022 Ranking," Brand Finance, accessed October 14, 2022, https://brandirectory.com/rankings/airlines/table.

8. James John Hoogerwerf, "From Crop Duster to Airline: The Origins of Delta Air Lines to World War II" (PhD diss., Auburn University, 2013), 20–21; James C. Giesen, *Boll Weevil Blues: Cotton, Myth, and Power in the American South* (Chicago: University of Chicago Press, 2011), 7–8, 48; Pete Daniel, *Breaking the Land: The Transformation of Cotton, Tobacco, and Rice Cultures since 1880* (Urbana: University of Illinois Press, 1986), 7–8.

9. Giesen, *Boll Weevil Blues*, ix–xi.

10. Giesen, *Boll Weevil Blues*, 79, 84.

11. John H. Perkins, *Insects, Experts, and the Insecticide Crisis: The Quest for New Pest Management Strategies* (New York: Plenum Press, 1982), 3–4; Giesen, *Boll Weevil Blues*,

86. For more on the use of arsenicals in agriculture, see Adam Romero, *Economic Poisoning: Industrial Waste and the Chemicalization of American Agriculture* (Oakland: University of California Press, 2021), 18–49.

12. Giesen, *Boll Weevil Blues*, 87–88; W. David Lewis and Wesley Phillips Newton, *Delta: The History of an Airline* (Athens: University of Georgia Press, 1979), 10–12. The first experiment with aerial crop dusting occurred in 1920 in the Sacramento Valley. At the time, California was leading the country in the development of capital-intensive, chemical-laden agriculture dependent on mechanized equipment. See Jack Temple Kirby, *Rural Worlds Lost: The American South, 1920–1960* (Baton Rouge: Louisiana State University Press, 1987), 11.

13. Lewis and Newton, *Delta*, 5–7, 11–12.

14. Kirby, *Rural Worlds Lost*, 53; David B. Danbom, *Born in the Country: A History of Rural America*, 3rd ed. (Baltimore: Johns Hopkins University Press, 2007), 119, 169; James C. Cobb, *The South and America since World War II* (New York: Oxford University Press, 2011), 52.

15. Kirby, *Rural Worlds Lost*, 34, 51–79. On the rise of factory farming in the 1910s and 1920s, see Deborah Fitzgerald, *Every Farm a Factory: The Industrial Ideal in American Agriculture* (New Haven, Conn.: Yale University Press, 2003).

16. Lewis and Newton, *Delta*, 13–14. In 1927, the company issued a pamphlet that declared, "Huff Daland Dusters, Inc., have dusted more acres of cotton by airplane than all other Agencies, including the U.S. Government, and have never failed to fulfill their contracts." *Airplane Dusting by Huff Daland Dusters, Inc. Returns Profits*, pamphlet, 1927, Huff Daland Dusters Records, box 1, folder 3, Delta Archives.

17. For excellent histories of the 1927 flood and its effects on the people and environment of the Mississippi Delta, see John M. Barry, *Rising Tide: The Great Mississippi Flood of 1927 and How It Changed America* (New York: Simon & Schuster, 1997); Pete Daniel, *Deep'n as It Come: The 1927 Mississippi River Flood* (Fayetteville: University of Arkansas Press, 1996); Hoogerwerf, "From Crop Duster to Airline," 105.

18. Lewis and Newton, *Delta*, 14, 17–19.

19. Lewis and Newton, *Delta*, 14, 17–21; "Delta Airlines: Women's History Month, Catherine Fitzgerald," MarketScreener, March 9, 2016, https://www.marketscreener.com/quote/stock/DELTA-AIR-LINES-INC-49285/news/Delta-Air-Lines-Women-s-History-Month-Catherine-Fitzgerald-By-Taking-Off--21987441/; Recorded Articles of Incorporation of Delta Air Services, Inc., December 3, 1928, Delta Air Service Records, box 1, folder 1, Delta Archives.

20. Hoogerwerf, "From Crop Duster to Airline," 164, 167; Lewis and Newton, *Delta*, 23.

21. Lewis and Newton, *Delta*, 30–44. The boll weevil continued to be a revenue generator for Delta. In a 1935 letter to C. E. Woolman, Bert E. Coad welcomed publicity in Dallas about boll weevil infestations, believing that pest fears might lead to more business. Letter from Bert E. Coad to C. E. Woolman, August 1935, C. E. Woolman Collection, box 2, folder 2, Delta Archives.

22. Lewis and Newton, *Delta*, 28–29; 1941 Delta Air Corporation Annual Report, n.p. Delta's general traffic manager in Atlanta, Laigh C. Parker, offered several accounts of early passenger operations that give some sense of what the business looked like at its inception. In July 1934, just days before Delta's passenger service got underway in

Texas, Parker spoke of last-minute efforts in Dallas to have "some sort of sign painted which will be put up over the front of the ticket window." He also scrambled to get "1506 gum and cotton packets from Wm. K. Wrigley," saying, "This should have been attended to, but it just slipped my mind." These were hectic times. A few months later, Parker explained that "business continues to be good over this way, Columbia having sold nine passengers in a space of five days, and Atlanta having already made bookings for space as far as November fifteenth." These numbers might seem paltry in the present-day era of high-volume sales made possible by online vendors such as Kayak, Orbitz, and Travelocity, but in the early years, nine ticket sales for one city was a good week's worth of work. Letter from Laigh C. Parker to C. E. Woolman, July 26, 1934; letter from Laigh C. Parker to C. E. Woolman, Monroe [La.], October 30, 1934, C. E. Woolman Collection, box 2, folder 9, Delta Archives.

23. Lewis and Newton, *Delta*, 83; 1941 Delta Air Corporation Annual Report, n.p.; 1945 Delta Air Corporation Annual Report, n.p.

24. Lewis and Newton, *Delta*, 106; 1945 Delta Air Corporation Annual Report, n.p.; Barbara A. Gilchrest, "Sun Exposure and Vitamin D Sufficiency," *American Journal of Clinical Nutrition* 88 (2008): 571Sl; Hector F. Deluca, "History of the Discovery of Vitamin D and Its Active Metabolites," *BoneKEy Reports* 3, no. 479 (2014): 1. The phrase "rural worlds lost" was southern historian Jack Temple Kirby's description of the depopulation of the rural South in the twentieth century, but it speaks to a larger pattern of urbanization that reshaped the United States in the twentieth century, in the North and South. See Kirby, *Rural Worlds Lost*. On Sunkist and Sun-Maid, see Ted Steinberg, *Down to Earth: Nature's Role in American History* (New York: Oxford University Press, 2002), 180–83. For an example of Delta's sunbathing advertising, see 1955 Delta Air Lines Annual Report, 20; 1956 Delta Air Lines Annual Report, 20.

25. Lewis and Newton, *Delta*, 83.

26. Raymond Arsenault, "The End of the Long Hot Summer: The Air Conditioner and Southern Culture," *Journal of Southern History* 50 (1984): 597–628. For a national history of air-conditioning, see Gail Cooper, *Air-Conditioning America: Engineers and the Controlled Environment, 1900–1960* (Baltimore: Johns Hopkins University Press, 1998). For a fascinating reanalysis of how air-conditioning reshaped the American South, including discussions of how this technology exacerbated racial inequalities in southern states, see Jason Hauser, "By Degree: A History of Heat in the Subtropical American South" (PhD diss., Mississippi State University, 2017), 232–85.

27. 1943 Delta Air Corporation Annual Report, 6.

28. For more on DDT deployment overseas and chemical companies' adaption to civilian markets after World War II, see Edmund P. Russell, *War and Nature: Fighting Humans and Insects with Chemicals from World War I to "Silent Spring"* (Cambridge: Cambridge University Press, 2001); David Kinkela, *DDT and the American Century: Global Health, Environmental Politics, and the Pesticide That Changed the World* (Chapel Hill: University of North Carolina Press, 2011). On the history of the American South as a diseased landscape, see Todd L. Savitt and James Harvey Young, eds., *Disease and Distinctiveness in the American South* (Knoxville: University of Tennessee Press, 1988). For the argument that New Deal programs helped to reduce malaria infection in the South, see Daniel Sledge and George Mohler, "Eliminating Malaria in the American South: An Analysis of the Decline of Malaria in 1930s Alabama," *American Journal of Public Health*

103, no. 8 (August 2013): 1381–92. For arguments about rural migration and malaria, see Margaret Humphreys, "How Four Once Common Diseases Were Eliminated from the American South," *Health Affairs* 28, no. 6 (2009): 1737. See also Margaret Humphreys, *Malaria: Poverty, Race, and Public Health in the United States* (Baltimore: Johns Hopkins University Press, 2001).

29. Anke Ortlepp, *Jim Crow Terminals: The Desegregation of American Airports* (Athens: University of Georgia Press, 2017), 92.

30. Ortlepp, *Jim Crow Terminals*, 13–18, 91–93.

31. Ortlepp, *Jim Crow Terminals*, 92–97.

32. Ortlepp, *Jim Crow Terminals*, 25.

33. "Delta Air Changes Name," *New York Times*, December 22, 1945, Business and Finance Section, 26; 1946 Delta Air Lines Annual Report, n.p.; 1952 Delta Air Lines Annual Report, 4, 16–17; 1953 Delta Air Lines Annual Report, 1; Lewis and Newton, *Delta*, 260–61; 1956 Delta Air Lines Annual Report, 1, 4.

34. For Delta flight paths in 1955, see Delta Air Lines System Timetable February 1, 1955, Delta Air Lines Serial Publication, Delta Flight Museum, https://deltamuseum.catalogaccess.com/; U.S. Census Bureau, *Census of Population: 1950*, vol. 1, *Number of Inhabitants* (Washington, D.C.: Government Printing Office, 1952), 13-9, 5-15, 18-13, 24-7.

CHAPTER 4

1. Mark Fiege, *Republic of Nature: An Environmental History of the United States* (Seattle: University of Washington Press, 2012), 372. For the types of gasoline Delta used in the 1930s, see letter from Shell Petroleum Company to Delta Air Corporation, June 8, 1935, C. E. Woolman Collection, box 2, folder 8, Delta Archives. For a discussion of the difference between motor gasoline and aviation gasoline at this time, see letter from R. C. Oertel, aviation department chief for Standard Oil of New Jersey, to C. E. Woolman, Delta Air Corporation, June 10, 1935, C. E. Woolman Collection, box 2, folder 8, Delta Archives. For a global environmental history that looks at petroleum production and consumption from the eighteenth century to today, see Brian Black, *Crude Reality: Petroleum in World History* (Lanham, Md.: Rowman and Littlefield, 2012).

2. Delta Air Corp's Gasoline Proposals, June 19, 1935; letter from Sinclair Refining Company to C. E. Woolman, June 11, 1935; letter from assistant sales manager of the Texas Company to Delta Air Corporation, June 6, 1935; letter from assistant district sales manager of Gulf Refining Company to Delta Air Corporation, June 8, 1935; letter from Shell Petroleum Corporation to Delta Air Corporation, June 8, 1935, C. E. Woolman Collection, box 2, folder 8, Delta Archives.

3. Per Month Basis, Two-Way Daily Service for Dallas-Birmingham, n.d., box 2, folder 10; letter from J. H. Doolittle of Shell Petroleum Corporation to C. E. Woolman, June 8, 1935, box 2, folder 8, Delta Archives.

4. Letter from R. C. Oertel to C. E. Woolman, June 10, 1935.

5. "Oil Industry Stepping Up Output of Aviation Fuel, Gasoline Rationing Due; More Well Drilling Needed," *Wall Street Journal*, January 5, 1942, 24; Delta Air Corporation Memorandum from Vice President and General Manager to All Delta Employees, May 31, 1943, C. E. Woolman Collection, box 1, folder 2, Delta Archives.

6. W. David Lewis and Wesley Phillips Newton, *Delta: The History of an Airline* (Athens: University of Georgia Press, 1979), 83, 87; James C. Cobb, *The South and America since World War II* (New York: Oxford University Press, 2011), 55; Gavin Wright, *Sharing the Prize: The Economics of the Civil Rights Revolution in the American South* (Cambridge, Mass.: Belknap Press, 2013), 24; 1944 Delta Air Lines Annual Report, 3; 1941 Delta Air Lines Annual Report, n.p.

7. "Oil Industry Stepping Up Output."

8. 1946 Delta Air Lines Annual Report, n.p.; 1948 Delta Air Lines Annual Report, n.p.; 1951 Delta Air Lines Annual Report, 3; 1952 Delta Air Lines Annual Report, 3; 1955 Delta Air Lines Annual Report, 3; 1956 Delta Air Lines Annual Report, 3; 1957 Delta Air Lines Annual Report, 2.

9. Lewis and Newton, *Delta*, 277; 1956 Delta Air Lines Annual Report, 5; 1959 Delta Air Lines Annual Report, 6. On the history of jet airlines and gas turbines, see Vaclav Smil, *Prime Movers of Globalization: The History and Impact of Diesel Engines and Gas Turbines* (Cambridge, Mass.: MIT Press, 2010), 79–108.

10. "Southern Airline Services Downturn," *Chapel Hill (N.C.) News*, August 29, 1982, D1, D7; 1987 Delta Air Lines Annual Report, 4; "1955," Delta News Hub, Delta Air Lines, March 19, 2013, https://news.delta.com/1955.

11. 1987 Delta Air Lines Annual Report, 4. Twenty years after adopting the hub-and-spoke system, Delta continued to stress the importance of smaller markets to its bottom line. Company managers said they remained committed to "small city service" because it enabled the firm "to serve all kinds of traffic—short-haul, short-haul connecting, and long-haul local and connecting." The company also claimed that it had "the shortest average aircraft hop and the shortest passenger haul" of all the major airlines. See House Subcommittee on Aviation, Committee on Public Works and Transportation, *Reform of the Economic Regulation of Air Carriers*, 94th Cong., 2nd sess., May 18, 1976, 393. In time, especially after Congress passed legislation deregulating the airline industry in the 1970s, most other major carriers would adopt Delta's hub-and-spoke flight network, but many people in the industry still acknowledged that Delta had been a trailblazer in developing the model. "Using local markets to feed long-haul service," as one observer put it, "is the guts of the most profitable operations in the industry today," noting that Delta was the "classic example." Senate Subcommittee on Aviation, Committee on Commerce, Science, and Transportation, *Impact of Airline Deregulation*, 96th Cong., 1st sess., April 25 and 27, 1979, 208.

12. 1961 Delta Air Lines Annual Report, 3.

13. Lewis and Newton, *Delta*, 326; "Transportation to Be Revolutionized by Jets," *Alton (Ill.) Evening Telegraph*, August 21, 1958, 26; Harry Lawrence, *Aviation and the Role of Government* (Dubuque, Iowa: Kendall Hunt, 2014), 151; Smil, *Prime Movers of Globalization*, 23, 34, 36, 39.

14. Energy historian Christopher Jones penned an excellent work showing how pipelines fueled expansion for many American industries in the twentieth century. See Christopher Jones, *Routes of Power: Energy and America* (Cambridge, Mass.: Harvard University Press, 2004); "Capital's Underground Is a 6900-Mile Pipe Dream," *Washington Post*, July 12, 1964, K5; "Colonial Pipeline Will Expand," *Atlanta Constitution*, February 2, 1971, 12A; "FTC Chairman Says the Takeover Could Have Crimped Atlanta Firm," *Atlanta Constitution*, August 10, 1982, 1C.

15. Anke Ortlepp, *Jim Crow Terminals: The Desegregation of American Airports* (Athens: University of Georgia Press, 2017), 9, 41, 42, 48–50, 134.

16. Ortlepp, *Jim Crow Terminals*, 27, 31, 70.

17. Gavin Wright, *Sharing the Prize: The Economics of the Civil Rights Revolution in the American South* (Cambridge, Mass.: Belknap Press of Harvard University Press, 2013), 18–19, 27, 74, 259; Edward L. Ayers, *Southern Journey: The Migrations of the American South, 1790–2020* (Baton Rouge: Louisiana State University Press, 2020), 93; Cobb, *The South and America*, 158.

18. 1974 Delta Air Lines Annual Report, 14. For more on fuel price fluctuations, see Clinton V. Oster Jr., "Impact of Rising Fuel Prices," in *Airline Deregulation: The Early Experience*, ed. John R. Meyer, Clinton V. Oster Jr., Ivor P. Morgan, Benjamin A. Berman, and Diana L. Strassman (Boston: Auburn House, 1981), 161–88.

19. 1974 Delta Air Lines Annual Report, 3–4, 14.

20. Lewis and Newton, *Delta*, 409; 1978 Delta Air Lines Annual Report, 5; 1979 Delta Air Lines Annual Report, 3.

21. 1975 Delta Air Lines Annual Report, 12; 1979 Delta Air Lines Annual Report, 6, 9.

22. Harvard faculty completed an excellent study on this topic in 1981. See Meyer et al., *Airline Deregulation*, 3–4.

23. 1980 Delta Air Lines Annual Report, 7, 9; 1981 Delta Air Lines Annual Report, 3, 10.

24. 1987 Delta Air Lines Annual Report, 1, 6; 1988 Delta Air Lines Annual Report, 1; 1983 Delta Air Lines Annual Report, 1; 1989 Delta Air Lines Annual Report, 1.

25. Joe Sharkey, "24 Small Towns May Lose Air Service," *New York Times*, July 19, 2011, B1; Senate Subcommittee on Aviation, Committee on Commerce, Science, and Transportation, *Impact of Airline Deregulation*, 96th Cong., 1st sess., April 25 and 27, 1979, 57–58; J. Richard Jones, "Twenty Years of Airline Deregulation: The Impact on Outlying Small Communities," *Journal of Transportation Management* (Fall 1998): 33–43; Senate Subcommittee on Rural Economy and Family Farming, Committee on Small Business, *The Effect of Airline Deregulation on the Rural Economy*, 100th Cong., 1st sess., October 28, 1987, 33.

26. Jeff Bailey, "Subsidies Keep Airlines Flying to Small Towns," *New York Times*, October 6, 2006, www.nytimes.com/2006/10/06/business/06boonies.html; Adam Hochberg, "Federal Subsidies Keep Small-Town Airports Flying," *Morning Edition*, NPR, November 18, 2009, www.npr.org/2009/11/18/120126620/federal-subsidies-keep-small-town-airports-flying; Geoff Jones, *Delta Air Lines: 75 Years of Airline Excellence* (Charleston, S.C.: Arcadia, 2003), 91.

27. Erin Moriarty, "Flying High Internationally," *Atlanta Business Chronicle*, November 15, 1999, www.bizjournals.com/atlanta/stories/1999/11/15/story7.html.

28. Ayers, *Southern Journey*, 104, 107; Migration Policy Institute, *The Nigerian Diaspora in the United States*, June 2015, 1–2, 4, https://www.migrationpolicy.org/sites/default/files/publications/RAD-Nigeria.pdf.

29. 2001 Delta Air Lines Annual Report, 10; 2004 Delta Air Lines Annual Report, 18.

30. House Committee on Transportation and Infrastructure, Subcommittee on Aviation, *Commercial Jet Fuel Supply: Impact and Cost on the United States Airline Industry*, 109th Cong., 2nd sess., 2006, 2; "America's Airlines, Flying on Empty," *The*

Economist, September 16, 2005, https://www.economist.com/unknown/2005/09/16/americas-airlines-flying-on-empty; "Will Katrina Ground Airlines for Good?," *Forbes*, September 1, 2005, https://www.forbes.com/2005/09/01/katrina-airlines-fuel-cz_mt_pb_0901airlines.html; Eduardo Porter, "Hurricane Katrina: Economic Impact," *New York Times*, August 31, 2005, C1.

31. "Delta Buys Oil Refinery," *News-Star* (Monroe, La.), May 1, 2012, 6C; Andrew Maykutch, "How (and Why) Delta Got into the Refinery Business," *Philadelphia Inquirer*, May 2, 2012, A1, A16; Elizabeth Caminiti, "Fueling the Climb: Spotlight on Jet Fuel Strategy," Delta News Hub, September 2, 2015, http://news.delta.com/fueling-climb-spotlight-jet-fuel-strategy.

32. Maykutch, "How (and Why) Delta"; Kate Kelly, "Delta Eyeing ConocoPhillips Refinery for More than $100 Million," CNBC, April 4, 2012, www.cnbc.com/; Delta Air Lines, "Delta Subsidiary to Acquire Trainer Refinery Complex" (press release), April 30, 2012, https://news.delta.com/delta-subsidiary-acquire-trainer-refinery-complex.

33. Steve Schaefer, "Why Buying a Refinery Could Be a Disaster for Delta Air Lines (Even with JPMorgan's Help)," *Forbes*, April 11, 2012, www.forbes.com/.

34. Schaefer, "Why Buying a Refinery."

35. Jarrett Renshaw, "Delta's Refinery Sacrifices Profits for Lower Fuel Cost—Memo," Reuters, August 3, 2016, www.reuters.com/.

36. Clive Irving, "Forget Futuristic Green Jets, You Could Be Flying Old Gas Guzzlers for Years to Come," *Daily Beast*, May 1, 2016, www.thedailybeast.com/forget-futuristic-green-jets-you-could-be-flying-old-gas-guzzlers-for-years-to-come.

37. Irving, "Forget Futuristic Green Jets"; Scott McCartney, "A Prius with Wings vs. a Guzzler in the Clouds," *Wall Street Journal*, August 12, 2010. Efficiency statistics calculated from analysis of fuel use in Delta Air Lines Annual Reports, 1980–88 and 2008–16.

38. 2016 Delta Air Lines Annual Report, 26.

39. 2016 Delta Air Lines Corporate Responsibility Report, 7, 37.

40. 2019 Delta Air Lines 10-K filing, 54; Delta Air Lines, "Delta's Corporate Responsibility Report Pledges Continued Commitment to Social Impact and Sustainability in Face of Pandemic" (press release), July 31, 2020, https://news.delta.com.

41. 2021 Delta Air Lines 10-K filing, 68; Niraj Chokshi, "Delta Reports Quarterly Loss, but Sees Better Days Ahead," *New York Times*, April 13, 2022, B6; "Airline and National Security Relief Programs," U.S. Department of the Treasury, accessed October 4, 2022, https://home.treasury.gov/policy-issues/coronavirus/assistance-for-american-industry/airline-and-national-security-relief-programs.

42. "Our Flight to Net Zero," Delta Air Lines, accessed November 1, 2022, www.delta.com/us/en/about-delta/sustainability; Siddharth Vikram Philip and Ben Elgin, "Airlines Rush toward Sustainable Fuel but Supplies Are Limited," *Bloomberg*, November 10, 2021, www.bloomberg.com/news/articles/2021-11-10/airlines-rush-toward-sustainable-fuel-but-supplies-are-limited.

43. Chokshi, "Delta Reports Quarterly Loss."

44. Delta Air Lines, "Delta to Suspend Flying in Select U.S. Cities" (press release), Delta News Hub, June 5, 2020, https://news.delta.com/delta-suspend-flying-select-us-cities; Sharkey, "24 Small Towns."

45. Delta Air Lines, "Delta to Suspend Flying"; U.S. Census Bureau, *City and Town Population Totals, 2010–2020*, www.census.gov.

CHAPTER 5

1. Sam Walton with John Huey, *Sam Walton: Made in America—My Story* (New York: Doubleday, 1992), 102–3; Lou Pritchett, *Stop Paddling and Start Rocking the Boat: Business Lessons from the School of Hard Knocks* (1995; repr., New York: Authors Choice Press, 2007), 27; Richard Feloni, "How the Walmart Shareholders Meeting Went from a Few Guys in a Coffee Shop to a 14,000-Person, Star-Studded Celebration," *Business Insider*, June 2, 2017, www.businessinsider.com/history-walmart-shareholders -meeting-2017-6. On Alice Walton's enthusiasm for the outdoors, see Bob Ortega, *In Sam We Trust: The Untold Story of Sam Walton and How Wal-Mart Is Devouring America* (New York: Times Books, 1998), 24; Walton, *Sam Walton*, 19; Vance H. Trimble, *Sam Walton: The Inside Story of America's Richest Man* (New York: Penguin, 1990), 45, 50.

2. Ortega, *In Sam We Trust*, 190; Walton, *Sam Walton*, 102–3.

3. Walton, *Sam Walton*, 102–3.

4. Isadore Barmash, "The Hot Ticket in Retailing: Wal-Mart Is Making a Bundle by Bringing Big Discounts to the Sunbelt," *New York Times*, July 1, 1984, 4F; Bethany Moreton, *To Serve God and Wal-Mart: The Making of Christian Free Enterprise* (Cambridge, Mass.: Harvard University Press, 2009), 132. For more on the history of cultural representations of southern working-class whites as antimodern "hillbillies" throughout the twentieth century, see Anthony Harkins, *Hillbilly: A Cultural History of an American Icon* (New York: Oxford University Press, 2004).

5. Pritchett, *Stop Paddling*, 26–31.

6. Pritchett, *Stop Paddling*, 29–32.

7. Walton, *Sam Walton*, 3, 4; Ortega, *In Sam We Trust*, 18, 19; Trimble, *Sam Walton*, 16. For more on the economic and ecological costs of the Dust Bowl, see Donald Worster, *Dust Bowl: The Southern Plains in the 1930s* (New York: Oxford University Press, 1979).

8. Michael Bergdahl, *What I Learned from Sam Walton: How to Compete and Thrive in a Wal-Mart World* (Hoboken, N.J.: John Wiley & Sons, 2004), 77; Bethany Moreton, "It Came from Bentonville: The Agrarian Origins of Wal-Mart Culture," in *Wal-Mart: The Face of Twenty-First-Century Capitalism*, ed. Nelson Lichtenstein (New York: New Press, 2006), 75; Moreton, *To Serve God and Wal-Mart*, 45.

9. Trimble, *Sam Walton*, 29–33.

10. Susan Strasser, "Woolworth to Wal-Mart: Mass Merchandising and the Changing Culture of Consumption," in Lichtenstein, *Wal-Mart*, 36; Sandra Stringer Vance and Roy Vernon Scott, *Wal-Mart: A History of Sam Walton's Retail Phenomenon* (New York: Twayne Publishers, 1994), 18; Daniel B. Schneider, "F.Y.I.," *New York Times*, April 13, 1997, www.nytimes.com/1997/04/13/nyregion/fyi-559601.html.

11. Richard Tedlow, *New and Improved: The Story of Mass Marketing in America* (New York: Basic Books, 1990), 293.

12. Strasser, "Woolworth to Wal-Mart," 45; Vance and Scott, *Wal-Mart*, 19; Ortega, *In Sam We Trust*, 39.

13. Vance and Scott, *Wal-Mart*, 20.

14. Ortega, *In Sam We Trust*, 23, 36; Strasser, "Woolworth to Wal-Mart," 43; Tedlow, *New and Improved*, 11, 259, 262–64, 270.

15. Susan Strasser, *Satisfaction Guaranteed: The Making of the American Mass Market* (New York: Pantheon Books, 1989), 81, 212–16; Richard Ohmann, *Selling Culture: Magazines, Markets, and Class at the Turn of the Century* (New York: Verso, 1996), 68–69; Wayne F. Fuller, "The South and the Rural Free Delivery of Mail," *Journal of Southern History* 25, no. 4 (November 1959): 499–521; William Cronon, *Nature's Metropolis: Chicago and the Great West* (New York: W. W. Norton, 1991), 333–40.

16. Ortega, *In Sam We Trust*, 38; Vance and Scott, *Wal-Mart*, 19; Strasser, "Woolworth to Wal-Mart," 45–47; Tedlow, *New and Improved*, 182, 189; Greg Thain and John Bradley, *Store Wars: When Walmart Comes to Town* (New York: Wiley, 2001), 7–8.

17. Strasser, "Woolworth to Wal-Mart," 51; Vance and Scott, *Wal-Mart*, 21–22; Ortega, *In Sam We Trust*, 35, 43; Thain and Bradley, *Store Wars*, 8; Shane Hamilton, *Supermarket USA: Food and Power in the Cold War Farms Race* (New Haven, Conn.: Yale University Press, 2018), 12–14. Hamilton's book offers an excellent examination of how Cold War politics shaped the expansion of supermarkets in the United States.

18. Tedlow, *New and Improved*, 214–15; Ortega, *In Sam We Trust*, 40–42; Moreton, *To Serve God and Wal-Mart*, 18, 64, 68.

19. Moreton, "It Came from Bentonville," 64, 68.

20. Trimble, *Sam Walton*, 34; Ortega, *In Sam We Trust*, 23.

21. Walton, *Sam Walton*, 17–18.

22. Walton, *Sam Walton*, 18–19; Frank Robson interview transcript, Voices of Oklahoma, Oklahoma Historical Society, n.d., 23, https://voicesofoklahoma.com/interviews/robson-frank/.

23. Trimble, *Sam Walton*, 37, 46; Ortega, *In Sam We Trust*, 9–10; Andy Serwer, "The Waltons: Inside America's Richest Family," *Fortune*, November 15, 2004, https://archive.fortune.com/.

24. Serwer, "The Waltons"; Frank Robson interview transcript, 3–4. On the Oklahoma oil boom, see Mark Fiege, *Republic of Nature: An Environmental History of the United States* (Seattle : University of Washington Press, 2012), 372.

25. Frank Robson interview transcript.

26. Vance and Scott, *Wal-Mart*, 5.

27. Walton, *Sam Walton*, 20, 22, 26; Moreton, *To Serve God and Wal-Mart*, 25; Trimble, *Sam Walton*, 46.

28. Walton, *Sam Walton*, 21; Vance and Scott, *Wal-Mart*, 5.

29. Walton, *Sam Walton*, 19, 21.

30. Walton, *Sam Walton*, 26–27.

31. Walton, *Sam Walton*, 25.

32. Trimble, *Sam Walton*, 55; Walton, *Sam Walton*, 30; Vance and Scott, *Wal-Mart*, 9.

33. Walton, *Sam Walton*, 30; Trimble, *Sam Walton*, 44, 47.

34. Trimble, *Sam Walton*, 58–59; Vance and Scott, *Wal-Mart*, 9; Ortega, *In Sam We Trust*, 29.

35. Walton, *Sam Walton*, 31; Bergdahl, *What I Learned from Sam Walton*, 133.

36. Walton, *Sam Walton*, 33.

37. Trimble, *Sam Walton*, 64, 71; Ortega, *In Sam We Trust*, 31; Moreton, *To Serve God and Wal-Mart*, 12; Vance and Scott, *Wal-Mart*, 13.

38. Walton, *Sam Walton*, 112.

39. Walton, *Sam Walton*, 40.

40. Don Soderquist, *The Wal-Mart Way: The Inside Story of the Success of the World's Largest Company* (Nashville: Nelson Business, 2005), 179; Walton, *Sam Walton*, 110.

41. William Graves, "Discounting Northern Capital: Financing the World's Largest Retailer from the Periphery," in *Wal-Mart World: The World's Biggest Corporation in the Global Economy*, ed. Stanley D. Brunn (New York: Routledge, 2006), 51.

42. Vance and Scott, *Wal-Mart*, 51; Moreton, *To Serve God and Wal-Mart*, 12.

43. House Committee on Public Works and Transportation, *To Examine the Future of the Nation's Infrastructure Needs*, 101st Cong., 2nd sess., March 29, June 13, July 16, August 7–8, 30–31, September 8, 17, December 7, 1990, 158.

44. Walton, *Sam Walton*, 111.

45. Moreton, "It Came from Bentonville," 63–64; Moreton, *To Serve God and Wal-Mart*, 35.

46. Vance and Scott, *Wal-Mart*, 14.

47. Thomas Jessen Adams, "Making the New Shop Floor: Wal-Mart, Labor Control, and the History of the Postwar Discount Retail Industry in America," in Lichtenstein, *Wal-Mart*, 214; Strasser, "Woolworth to Wal-Mart," 52; Vance and Scott, *Wal-Mart*, 45.

48. Walton, *Sam Walton*, 109; Trimble, *Sam Walton*, 83, 84.

49. Walton, *Sam Walton*, 42, 45.

50. Walton, *Sam Walton*, 43, 50.

51. Walton, *Sam Walton*, 43, 45–46; Trimble, *Sam Walton*, 101–2.

52. Walton, *Sam Walton*, 50, 56; Ortega, *In Sam We Trust*, 57; Vance and Scott, *Wal-Mart*, 48; Trimble, *Sam Walton*, 125.

53. Ortega, *In Sam We Trust*, 58.

54. Vance and Scott, *Wal-Mart*, 69; Moreton, *To Serve God and Wal-Mart*, 28.

55. Nelson Lichtenstein, "Wal-Mart: A Template for Twenty-First-Century Capitalism," in Lichtenstein, *Wal-Mart*, 14; Moreton, *To Serve God and Wal-Mart*, 11.

56. Moreton, *To Serve God and Wal-Mart*, 84.

57. Moreton, *To Serve God and Wal-Mart*, 39; James C. Cobb, *The Selling of the South: The Southern Crusade for Industrial Development, 1936–1990* (Urbana: University of Illinois Press, 1993), 101.

58. Trimble, *Sam Walton*, 229; Ortega, *In Sam We Trust*, 87.

59. Edward Humes, *Force of Nature: The Unlikely Story of Wal-Mart's Green Revolution* (New York: Harper Business, 2011), 37–38; Nelson Lichtenstein, *The Retail Revolution: How Wal-Mart Created a Brave New World of Business* (New York: Picador, 2010), 117–18.

60. Humes, *Force of Nature*, 39–40.

61. Moreton, "It Came from Bentonville," 82; Moreton, *To Serve God and Wal-Mart*, 71, 102.

62. On long-haul trucking and Walmart, see Shane Hamilton, *Trucking Country: The Road to America's Wal-Mart Economy* (Princeton, N.J.: Princeton University Press, 2008).

63. Hamilton, *Trucking Country*.

64. Walton, *Sam Walton*, 216.

65. Vance and Scott, *Wal-Mart*, 47; Walton, *Sam Walton*, 93.

66. Walton, *Sam Walton*, 92–99.

67. Walton, *Sam Walton*, 93–97.

68. Trimble, *Sam Walton*, 139, 157, 161.

69. Walton, *Sam Walton*, 207–8; Humes, *Force of Nature*, 42.

70. Trimble, *Sam Walton*, 195; Vance, *Wal-Mart*, 84, 93–95, 115–21; Charles Fishman, *The Wal-Mart Effect: How the World's Most Powerful Company Really Works—and How It's Transforming the American Economy* (New York: Penguin, 2006), 75; Bergdahl, *What I Learned from Sam Walton*, 3.

71. Walton, *Sam Walton*, 90.

CHAPTER 6

1. Malcolm Gladwell, "Wal-Mart Encounters a Wall of Resistance in Vermont," *Washington Post*, July 27, 1994, A3; "City and Town Postcensal Tables: 1990–2000," U.S. Census Bureau, accessed October 31, 2022, www.census.gov.

2. "Williston Project," *Burlington Free Press*, January 18, 1992, 8A; "Greenfield Seeks Office in Williston," *Burlington Free Press*, January 19, 1993, B1.

3. For extensive records detailing WCRG's extended fight with Walmart, see Williston Citizens for Responsible Growth (WCRG) Records, 1988–1998, Vermont Historical Society, Barre, Vt.; "Small Town Worried by Big Retailer," *Daily Chronicle* (Dekalb, Ill.), January 30, 1997, 5A; "Greenfield Seeks Office in Williston," *Burlington Free Press*, January 19, 1993, B1; "Wal-Mart: Retailer Likely to Put Up a Tough Fight in NCCo," *News Journal* (Wilmington, Del.), March 9, 1997, A11.

4. For an excellent summary of WCRG history, see the Vermont Historical Society's WCRG finding aid, accessed October 16, 2022, https://vermonthistory.org/documents/findaid/WCRG.pdf; author interview with Fred "Chico" Lager, March 29, 2019.

5. Keith Henderson, "'Sprawl-Mart' Endangers Vermont," *Christian Science Monitor*, December 6, 1993.

6. Micah Cohen, "'New' Vermont Is Liberal, but 'Old' Vermont Is Still There," FiveThirtyEight, October 1, 2012, https://fivethirtyeight.com/features/new-vermont-is-liberal-but-old-vermont-is-still-there/.

7. James C. Cobb, *The South and America since World War II* (New York: Oxford University Press, 2011), 180–81. On the emergence of this new color-blind politics centered on consumer choice and individualism, see Matthew D. Lassiter, *Silent Majority: Suburban Politics in the Sunbelt South* (Princeton, N.J.: Princeton University Press, 2006); Lily Geismer, *Don't Blame Us: Suburban Liberals and the Transformation of the Democratic Party* (Princeton, N.J.: Princeton University Press, 2015); and Lisa McGirr, *Suburban Warriors: The Origins of the New American Right* (Princeton, N.J.: Princeton University Press, 2001).

8. Edward Humes, *Force of Nature: The Unlikely Story of Wal-Mart's Green Revolution* (New York: Harper Business, 2011), 41; Bob Ortega, *In Sam We Trust: The Untold Story of Sam Walton and How Wal-Mart Is Devouring America* (New York: Times Books, 1998), 215.

9. Vermont Environment District #4, Land Use Permit, dated April 27, 1988, doc. 787, folder 34, part 4; Vermont Environmental Board, Memorandum of Decision Re: Taft Corners Associates, Inc., Application #4c0696-11-EB, doc. 787, folder 1; memorandum from Chico Lager to Gerry Tarrant and Bob Morris, December 5, 1993, doc. 787, folder 34, part 2, WCRG Records.

10. Memorandum from Chico Lager to Gerry Tarrant and Bob Morris, December 5, 1993, Re: Key Points from the Original TCA Act 250 Permit, doc. 787, folder 34, part 2; memorandum from Chico Lager to Gerry Tarrant and Bob Morris, December 5, 1993, Re: Bob's Testimony—First Draft, doc. 787, folder 34, part 1, WCRG Records.

11. Memorandum from Chico Lager to Gerry Tarrant and Bob Morris, December 5, 1993, Key Points from the Original TCA Act 250 Permit, doc. 787, folder 34, part 2; memorandum from Chico Lager to Gerry Tarrant and Bob Morris, December 5, 1993, Re: Bob's Testimony—First Draft, doc. 787, folder 34, part 1, WCRG Records.

12. "Wal-Mart Gets OK, But . . . ," *Burlington Free Press*, July 30, 1994, 1A; "The Issues Slowly Grind through the Courts," *CRG [Citizens for Responsible Growth] Newsletter*, September 1993, 3, doc. 787, folder 7, part 2, WCRG Records.

13. "Endangered Label Was Wake-Up Call for State," *Burlington Free Press*, February 22, 2004, 1C.

14. "From the Co-presidents," *CRG Newsletter*, September 1993, 2, doc. 787, folder 7, part 2, WCRG Records.

15. "State Backs Retailer," *Burlington Free Press*, June 20, 1995, 1A; "Wal-Mart to Pave Own Way," *Burlington Free Press*, August 22, 1995, 1A, 14A; "Timeline: Wal-Mart in Vermont," *Burlington Free Press*, July 30, 1994, 5A; "Wal-Mart Wins in Court," *Burlington Free Press*, October 20, 1994, 1B.

16. "Wal-Mart to Pave Own Way."

17. "Wal-Mart to Pave Own Way."

18. "Wal-Mart: Retailer Hits Traffic Jam," *Burlington Free Press*, September 9, 1995, 4B.

19. "Wal-Mart Wins Final Approval," *Burlington Free Press*, October 19, 1995, 1A.

20. "Construction Vehicles Damaged at Wal-Mart," *Burlington Free Press*, October 31, 1995, 3B.

21. "Wal-Mart: Symbol or Just Another Store," *Burlington Free Press*, January 27, 1997, 1B, 2B.

22. "More Object to Retail Sprawl," *Burlington Free Press*, January 14, 1996, 1A.

23. "Wal-Mart: Store's Impact on Jobs Unclear," *Burlington Free Press*, October 21, 1996, 12A; "Wal-Mart: Symbol or Just Another Store."

24. Louis Hyman, *Debtor Nation: The History of America in Red Ink* (Princeton: Princeton University Press, 2011), 222.

25. Town of Willison Planning Commission, meeting minutes, June 16, 1987, 3–4, doc. 787, folder 6—part 1, WCRG Records.

26. Vermont Environmental Board, Memorandum of Understanding Re: Taft Corner Associates, Inc., Application #4C0696-11-EB, March 31, 1992, 8, doc. 787, folder 1, WCRG Records.

27. *United States of America v. Wal-Mart Stores, Inc.*, complaint, May 12, 2004, 6, https://www.epa.gov/sites/default/files/2013-09/documents/walmart2-cp.pdf.

28. *United States of America v. Wal-Mart Stores, Inc.*, 7, 12, 19, 23, 31.

29. *United States of America v. Wal-Mart Stores, Inc.*, 6.

30. "Wal-Mart II Clean Water Act," Environmental Protection Agency, www.epa.gov/enforcement/wal-mart-ii-clean-water-act-settlement.

31. Stone Environmental, *Town of Williston: Town-Wide Watershed Improvement Plan, Phase I*, Stone Project ID 12-055, February 28, 2013, 95, www.town.williston.vt.us; "Stream Monitoring: Muddy Brook," Rethink Runoff, accessed October 17, 2018,

http://rethinkrunoff.org/explore-the-lake-champlain-basin/muddy-brook/; Vermont Department of Environmental Conservation, Watershed Management Division, "2018 303 (d) List of Impaired Waters," September 2018, 5, https://dec.vermont.gov/sites/dec/files/documents/mp_PriorityWatersList_PartA_303d_2018.pdf.

32. Evan P. Fitzgerald and Samuel P. Parker, *Muddy Brook Phase 1 and 2 Stream Geomorphic Assessment Summary*, Fitzgerald Environmental Associates, February 2, 2009, 58, www.town.williston.vt.us.

33. Vermont District Environmental Commission #4, Application #4C0696, Findings of Fact and Conclusions of Law and Order, 10 VSA, Chapter 151 (Act 250), 4, doc. 787, folder 34, part 2, WCRG Records.

34. "Talking to Neighbors," *Burlington Free Press*, January 20, 1997, 6A; "Hearing on Retail Ban Divides Town," *Burlington Free Press*, May 9, 1997, 4B.

35. "Hearing on Retail Ban."

36. Isaac Olson, "Consultant Rolls Out Plan to Reduce Backups on I-89 Ramp," *Williston (Vt.) Observer*, June 1, 2005, www.willistonobserver.com/consultant-rolls-out-plan-to-reduce-backups-on-i-89-ramp/; "Vt. Lays Out 34 Highway Projects to Ease Chittenden County Traffic Congestion," NBC 5, November 26, 2013, www.mynbc5.com/.

37. To view the EPA's AirData Air Quality Monitors app, which shows the location of air quality monitors around the country as well as air quality data, see www.epa.gov/outdoor-air-quality-data/interactive-map-air-quality-monitors.

38. "Small Town Worried about Wal-Mart," *Daily Herald* (Tyrone, Pa.), January 30, 1997, 3; "Rural worlds lost" is southern historian Jack Temple Kirby's phrase. See Jack Temple Kirby, *Rural Worlds Lost: The American South, 1920–1960* (Baton Rouge: Louisiana State University Press, 1987).

39. Kenneth Stone, "The Effect of Wal-Mart Stores on Businesses in Host Towns and Surrounding Towns in Iowa," November 9, 1988, www2.econ.iastate.edu/faculty/stone/; Kenneth E. Stone, "Impact of the Wal-Mart Phenomenon on Rural Communities," in *Increasing Understanding of Public Problems and Policies 1997* (Chicago: Farm Foundation, 1997), 2, www2.econ.iastate.edu/faculty/stone/. For the updated study that highlights how Walmart stores in some markets actually increased "host-town retail sales," see Georgeanne M. Artz and Kenneth E. Stone, "Revisiting WalMart's Impact on Iowa Small-Town Retail: 25 Years Later," *Economic Development Quarterly* 26, no. 4 (October 2012): 298–310. For an archival collection containing Kenneth Stone's research materials, see the Kenneth Stone Papers, 1978–1999, Parks Library, Iowa State University, Ames. David Neumark, Junfu Zhang, and Stephen Ciccarella, "The Effects of Wal-Mart on Local Labor Markets," *Journal of Urban Economics* 63, no. 2 (March 2008): 405–30; Emek Basker, "Job Creation or Destruction? Labor Market Effect of Wal-Mart Expansion," *Review of Economics and Statistics* 87, no. 1 (February 2005): 174–83; Holly R. Barcus, "Wal-Mart-Scapes in Rural and Small-Town America," in *Wal-Mart World: The World's Biggest Corporation in the Global Economy*, ed. Stanley D. Brunn (New York: Routledge, 2006), 69; David Karjanen, "The Wal-Mart Effect and the New Face of Capitalism: Labor Market and Community Impacts of the Megaretailer," in *Wal-Mart: The Face of Twenty-First-Century Capitalism*, ed. Nelson Lichtenstein (New York: New Press, 2006), 150; Charles Fishman, *The Wal-Mart Effect: How the World's Most Powerful Company Really Works—and How It's Transforming the American Economy* (New York: Penguin, 2006), 143.

40. Letter from Bonnie Canning-Hofmann, president of Otsego County Conservation Association, to H. Lee Scott, vice president of Wal-Mart, December 14, 1994, series: Other Projects and Issues, 1969–2001, box 2, folder 2, OCCA Records.

41. Letter from Canning-Hofmann to Scott, December 14, 1994; letter from H. Lee Scott to Bonnie Canning-Hofmann, January 6, 1995, series: Other Projects and Issues, 1969–2001, box 2, folder 2, OCCA Records.

42. "Wal-Mart Cancels Store Plans after Residents Complain," *Rutland (Vt.) Daily Herald*, September 17, 1993, B8; "In 2 Towns, Main Street Fights Off Wal-Mart," *New York Times*, October 21, 1993, A16; "Wal-Mart Ready for Opponents, Right Down to Saving the Frogs," *Rutland (Vt.) Daily Herald*, September 4, 1993, 11.

43. "Wal-Mart, Builder Drop Plan for Steamboat Springs Store," *Daily Sentinel* (Grand Junction, Colo.), December 3, 1987, 3A; "Letters from Boynton Middle School, Wal-Mart Foes and a Journal Critic," *Ithaca (N.Y.) Journal*, May 14, 1994, 14A; "2 Communities Rejected Wal-Mart Developments," *Jackson Hole (Wyo.) Guide*, April 26, 1989, A6; "City Panel Challenges Market Study," *Iowa City Press-Citizen*, March 14, 1989, 1B; Ortega, *In Sam We Trust*, 171–72, 176, 180, 186, 300; Sandra S. Vance and Roy V. Scott, *Wal-Mart: A History of Sam Walton's Retail Phenomenon* (New York: Twayne Publishers, 1994), 153.

44. Ortega, *In Sam We Trust*, 298–300.

45. Paul Ingram and Lori Qingyuan Yue, "Trouble in Store: Probes, Protests, and Store Openings by Wal-Mart, 1998–2007," *American Journal of Sociology* 116, no. 1 (July 2010): 53; Don Soderquist, *The Wal-Mart Way: The Inside Story of the Success of the World's Largest Company* (Nashville: Nelson Business, 201); Matthew A Zook and Mark Graham, "Wal-Mart Nation: Mapping the Reach of a Retail Colossus," in Brunn, *Wal-Mart World*, 15–16.

46. Bethany Moreton, *To Serve God and Wal-Mart: The Making of Christian Free Enterprise* (Cambridge, Mass.: Harvard University Press, 2009), 251, 252; Ortega, *In Sam We Trust*, 205; Peter T. Kilborn, "Walmart's 'Buy American,'" *New York Times*, April 10, 1985, D1.

47. Ortega, *In Sam We Trust*, 223–24; Michael Bergdahl, *What I Learned from Sam Walton: How to Compete and Thrive in a Wal-Mart World* (Hoboken, N.J.: John Wiley & Sons, 2004), 16; Moreton, *To Serve God and Wal-Mart*, 253.

48. Moreton, *To Serve God and Wal-Mart*, 251.

49. Zook and Graham, "Wal-Mart Nation," 15.

50. Zook and Graham, "Wal-Mart Nation," 15–16.

51. Walton, *Sam Walton*, 204.

52. Zook and Graham, "Wal-Mart Nation," 16; Steve Burt and Leigh Sparks, "Wal-Mart's World," in Brunn, *Wal-Mart World*, 33–37.

53. Yuko Aoyama and Guido Schwarz, "The Myth of Wal-Martization: Retail Globalization and Local Competition in Japan and Germany," in Brunn, *Wal-Mart World*, 282–83; Mark Landler, "Wal-Mart to Abandon Germany," *New York Times*, July 29, 2006, B1.

54. Peter J. Hugill, "The Geostrategy of Global Business: Wal-Mart and Its Historical Forbears," in Brunn, *Wal-Mart World*, 12; Aoyama and Schwarz, "Myth of Wal-Martization," 289–91; Xue Hong, "Outsourcing in China: Walmart and Chinese Manufacturers," in *Walmart in China*, ed. Anita Chan (Ithaca, N.Y.: Cornell University Press, 2011), 36.

55. Zook and Graham, "Wal-Mart Nation," 15.

56. 2018 Walmart Annual Report, 23.

57. For more on "the squeeze," see Fishman, *Wal-Mart Effect*, 79–109.

58. Edna Bonacich with Khaleelah Hardie, "Wal-Mart and the Logistics Revolution," in Lichtenstein, *Wal-Mart*, 176; Jake Rosenfeld, *You're Paid What You're Worth: And Other Myths of the Modern Economy* (Cambridge, Mass.: Harvard University Press, 2021), 117.

59. Bonacich with Hardie, "Wal-Mart and the Logistics Revolution," 180.

60. Jessica F. Green, "Why Do We Need New Rules on Shipping Emissions? Well, 90 Percent of Global Trade Depends on Ships," *Washington Post*, April 17, 2018, www.washingtonpost.com/news/monkey-cage/wp/2018/04/17/why-do-we-need-new-rules-on-shipping-emissions-well-90-of-global-trade-depends-on-ships/; Naomi Klein, *This Changes Everything: Capitalism vs. the Climate* (New York: Simon and Schuster, 2014), 79. On HFO, see Vaclav Smil, *Prime Movers of Globalization: The History and Impact of Diesel Engines and Gas Turbines* (Cambridge, Mass.: MIT Press, 2010), 35.

61. 2018 Walmart Global Responsibility Report, 127; 2015 Walmart Global Responsibility Report, 100; Lichtenstein, "Walmart's Long March to China," 15.

62. Humes, *Force of Nature*, 21, 25–26.

63. Humes, *Force of Nature*, 53.

64. Moreton, *To Serve God and Wal-Mart*, 49, 253; Ortega, *In Sam We Trust*, 224–27, 331, 341, 345; Humes, *Force of Nature*, 43–51; Lee S. VanderVelde, "Wal-Mart as a Phenomenon in the Legal World: Matters of Scale, Scale Matters," in Brunn, *Wal-Mart World*, 126.

65. Humes, *Force of Nature*, 49; U.S. Justice Department, "Wal-Mart to Pay $400,000 Penalty and Cease Sales of Ozone-Depleting Refrigerants" (press release), January 22, 2004, https://www.justice.gov/archive/opa/pr/2004/January/04_enrd_040.htm; Margath A. Walker, David Walker, and Yanga Villagómez Velázquez, "The Wal-Martification of Teotihuacán: Issues of Resistance and Cultural Heritage," in Brunn, *Wal-Mart World*, 218; James J. Biles, "Globalization of Food Retailing and the Consequences of Wal-Martization in Mexico," in Brunn, *Wal-Mart World*, 349.

66. Humes, *Force of Nature*, 52.

67. Humes, *Force of Nature*, 53.

68. Humes, *Force of Nature*, 75.

69. On Jib Ellison and his first encounters with Walmart, see Humes, *Force of Nature*, 15–23, 55–70.

70. Humes, *Force of Nature*, 22–23, 74–75.

71. Humes, *Force of Nature*, 75–96.

72. H. Lee Scott, "Twenty-First Century Leadership," speech, Bentonville, Ark., October 25, 2005, https://corporate.walmart.com/twenty-first-century-leadership; Humes, *Force of Nature*, 100–104.

73. 2015 Walmart Global Responsibility Report, 5; Michael Barbaro, "Wal-Mart Puts Some Muscle behind Power-Sipping Bulbs," *New York Times*, January 2, 2007, A1; Humes, *Force of Nature*, 87–89, 141.

74. Humes, *Force of Nature*, 84, 132–34, 159–80.

75. "THESIS Index," Walmart Sustainability Hub, accessed October 17, 2022, www.walmartsustainabilityhub.com/sustainability-index; "THESIS the Sustainability Insight System," Sustainability Consortium, accessed October 17, 2022, https://sustainabilityconsortium.org/thesis/; Humes, *Force of Nature*, 182–201.

76. 2017 Walmart Global Responsibility Report, 7, 52, 54; 2018 Walmart Global Responsibility Report, 3; "Walmart Makes Bold Climate Commitments—and Delivers," We Are Still In, accessed October 17, 2022, www.wearestillin.com/success /walmart-makes-bold-climate-commitments-and-delivers.

77. Orville Schell, "How Walmart Is Changing China," *The Atlantic*, December 2011, www.theatlantic.com/magazine/archive/2011/12/how-walmart-is-changing-china /308709/; 2011 Walmart Global Responsibility Report, 101.

78. Walmart, climate change report for Carbon Disclosure Project, 2018, 37–38, accessed October 17, 2022, www.cdp.net/en.

79. Stacy Mitchell, "Walmart's Assault on the Climate: The Truth behind One of the Biggest Climate Polluters and Slickest Greenwashers in America," Institute for Local Self-Reliance, November 2013, 7, https://ilsr.org/wp-content/uploads/2013/10/ILSR -_Report_WalmartClimateChange.pdf; Stacy Mitchell, "The Truth behind Walmart's Green Claims," *HuffPost*, June 1, 2014, www.huffpost.com/entry/walmart-climate -change_b_5063035. Stacy Mitchell is the author of *Big-Box Swindle: The True Cost of Mega-Retailers and the Fight for America's Independent Businesses* (Boston: Beacon, 2006). In the company's 2018 CDP report, Walmart said it was able to "estimate the emissions from our third-party logistics coordinators in *some* of our markets using EPA emission factors for fuels in 2015" (emphasis added). This apparently included rough calculations for some container ships, but reporting here was vague at best. See Walmart 's 2018 climate change report for the CDP.

80. 2015 Walmart Global Responsibility Report, 55; 2017 Walmart Global Responsibility Report, 55.

81. "ESG Commitments and Progress," Walmart corporate website, accessed October 17, 2022, https://corporate.walmart.com/esgreport/reporting-data/esg -commitments-progress.

82. "THESIS Index," Walmart Sustainability Hub, accessed October 17, 2022, www .walmartsustainabilityhub.com/sustainability-index.

83. On the history of the Republican Party's former embrace of environmental policies and conservatives' subsequent abandonment of conservation politics, see Gregg Coodley and David Sarasohn, *The Green Years: When Democrats and Republicans United to Repair the Earth* (Lawrence: University Press of Kansas, 2021).

84. Bret Wallach, *A World Made for Money: Economy, Geography, and the Way We Live Today* (Lincoln: University of Nebraska Press, 2014), 19.

CHAPTER 7

1. Dave Hirschman, "'Cast Away' Delivers Goods for FedEx," *Chicago Tribune*, January 8, 2001, section 5, 2.

2. John Lippman and Rick Brooks, "FedEx Has a Star Turn in New 'Cast Away,'" *Wall Street Journal*, December 11, 2000, www.wsj.com/.

3. Hirschman, "'Cast Away' Delivers Goods."

4. Hirschman, "'Cast Away' Delivers Goods."

5. On Fred Smith's crediting Delta for the hub-and-spoke aviation model, see "Fred Smith on the Birth of FedEx," *Bloomberg*, September 20, 2004, https://www.bloomberg .com/news/articles/2004-09-19/online-extra-fred-smith-on-the-birth-of-fedex; Art Carden, "Smith, Frederick W.," Tennessee Encyclopedia, Tennessee Historical Society,

last updated March 1, 2018, http://tnency.utk.tennessee.edu/entries/frederick-w-smith/; Clare Lyster, *Learning from Logistics: How Networks Change Our Cities* (Basel: Birkhäuser, 2016), 53. On FedEx's aggressive expansion in the Philippines, China, Indonesia, and Japan, see John T. Bowen Jr., Thomas R. Leinbach, and Daniel Mabazza, "Air Cargo Services, the State and Industrialization Strategies in the Philippines: The Redevelopment of Subic Bay," *Regional Studies* 36, no. 5 (2002): 451; Robert Franks, "U.S. Express Carriers Deliver the Goods as Asia Bounces Back," *Wall Street Journal*, December 21, 1999, A18; Associated Press, "New Japan Routes for Federal Express," *New York Times*, June 3, 1997, D4.

6. "Behind the Fast Rise by Federal Express," *Washington Post*, August 2, 1978, C1, C5; "Fly-by-Night Success," *Atlanta Constitution*, June 14, 1981, K20; Roger Frock, *Changing How the World Does Business: FedEx's Incredible Journey to Success—The Inside Story* (San Francisco: Berrett-Koehler, 2006), 10.

7. "Vietnam Experience Inspires Veteran to Create Overnight Delivery Company," U.S. Army, February 28, 2014, www.army.mil/article/121066/loombe_experience_inspires_veteran_to_create_overnight_delivery_company.

8. Frock, *Changing How the World Does Business*, 10; "Federal Express Wasn't an Overnight Success," *Wall Street Journal*, June 6, 1989, B2.

9. "Federal Express Wasn't an Overnight Success."

10. "Behind the Fast Rise by Federal Express," *Washington Post*, August 2, 1978, C1. On the legendary "C" grade, see his 2004 interview, "Fred Smith on the Birth of FedEx," *Bloomberg Businessweek*, September 20, 2014, www.bloomberg.com/news/articles/2004-09-19/online-extra-fred-smith-on-the-birth-of-fedex.

11. Frock, *Changing How the World Does Business*, 12–13.

12. Winston Williams, "Overnight Delivery: The Battle Begins," *New York Times*, January 7, 1979, F1; Frock, *Changing How the World Does Business*, 19–20; "Bumpier Flight," *Wall Street Journal*, October 23, 1989, A1.

13. Frock, *Changing How the World Does Business*, 15–16, 19–20.

14. "The Small Package Airline Heading for a Big Bundle," *Los Angeles Times*, October 8, 1975, D11; Frock, *Changing How the World Does Business*, 45.

15. Frock, *Changing How the World Does Business*, 48.

16. Frock, *Changing How the World Does Business*, 52–53.

17. Daniel Terdiman, "Welcome to 'the Matrix': At FedEx's Sorting Hub, 1 Night, 1.5M Packages," CNET, July 12, 2014, https://www.cnet.com/science/at-fedex-sorting-packages-1-5-million-at-a-time/.

18. Author interview with Richard Smith, FedEx regional president of the Americas and executive vice president, March 15, 2021; Frock, *Changing How the World Does Business*, 54.

19. Author interview with Richard Smith; Frock, *Changing How the World Does Business*, 54; "Small-Cargo Firm Flies into Black," *Chicago Tribune*, March 1, 1976, C7; 1982 Federal Express Annual Report, 5.

20. Frock, *Changing How the World Does Business*, 54.

21. Frock, *Changing How the World Does Business*, 28.

22. Wanda Rushing, *Memphis and the Paradox of Place: Globalization in the American South* (Chapel Hill: University of North Carolina Press, 2009), 10–13, 26, 34, 84, 88–89.

23. Rushing, *Paradox of Place*, 11–12, 84, 89; Janann Sherman, *Walking on Air: The Aerial Adventures of Phoebe Omlie* (Oxford: University Press of Mississippi, 2011), 45–46.

24. Frock, *Changing How the World Does Business*, 54.

25. "Fed Ex, Airport Negotiate on Center," *Commercial Appeal* (Memphis, Tenn.), September 3, 1989, C1.

26. Williams, "Overnight Delivery"; Rushing, *Paradox of Place*, 85.

27. Rushing, *Paradox of Place*, 98–104. For the case of RCA and why that firm looked to take advantage of Memphis's low-wage economy in the 1960s, see Jefferson Cowie, *Capital Moves: RCA's Seventy-Year Quest for Cheap Labor* (Ithaca, N.Y.: Cornell University Press, 2019), 73–99.

28. "Small-Cargo Firm Flies into Black."

29. Rushing, *Memphis and the Paradox of Place*, 114.

30. Rushing, *Memphis and the Paradox of Place*, 85, 96; James C. Cobb, *The Selling of the South: The Southern Crusade for Industrial Development, 1936–1990* (Urbana: University of Illinois Press, 1993).

31. Frock, *Changing How the World Does Business*, 79.

32. Frock, *Changing How the World Does Business*, 59–60, 79–80.

33. Frock, *Changing How the World Does Business*, 101.

34. Frock, *Changing How the World Does Business*, 108.

35. Frock, *Changing How the World Does Business*, 112–13.

36. "Federal Express Still Top Banana of Fly-by-Nights," *Washington Post*, October 31, 1982, F1; "U.P.S. Delivers Challenge," *New York Times*, September 25, 1982, 39.

37. Frock, *Changing How the World Does Business*, 114, 139; Robert D. Mefadden, "Some Progress Reported in Talks Between U.P.S. and Teamsters," *New York Times*, November 17, 1974, 64; "Overnight Delivery."

38. "Small-Cargo Firm Flies into Black."

39. "Small-Cargo Firm Flies into Black"; "Wall Street Has Its Eye on Federal Express," *Atlanta Constitution*, September 21, 1977, 4C; 1978 Federal Express Annual Report, 2; "Behind the Fast Rise by Federal Express," *Washington Post*, August 2, 1978, C1; Bruce G. Leto, "Deregulatory Scheme Had No Effect on the Applicable Substantive Law to Determine Liability of Shipper for Lost Shipment of Goods," *Villanova Law Review* 30, no. 3 (1985): 890.

40. "Wall Street Has Its Eye"; Williams, "Overnight Delivery"; "Overseas Restrictions Handicap Fed Ex Goals," *Commercial Appeal* (Memphis, Tenn.), September 17, 1990, B5; "Fed Ex Awaits Landing Rights in Five Nations," *Commercial Appeal*, July 30, 1989, C1, C2; Andrea Adelson, "Federal Express to Buy Flying Tiger," *New York Times*, December 17, 1988, 35, 37; "Federal Express Green Light to Start $100 M Business in Subic," *Philippine Star*, January 24, 1994, 21.

41. Adelson, "Federal Express to Buy." For the story of how an air service built by some of the original Flying Tiger airmen from World War II later participated in covert operations during the Vietnam War, see Richard Halloran, "Air America's Civilian Façade Gives It Latitude in East Asia," *New York Times*, April 5, 1970, 1; "Federal Express Needs a Lift Overseas after Expansion," *Wall Street Journal*, March 22, 1989, A9.

42. "Federal Express Needs a Lift."

43. 1990 Federal Express Annual Report, 23–24, 40; "Federal Express Needs a Lift."

44. 1992 Federal Express Annual Report, 4.

45. "Turbulence Ahead: Federal Express Faces Challenges to Its Grip on Overnight Delivery," *Wall Street Journal*, January 8, 1988, 1.

46. 1990 Federal Express Annual Report, front matter.

CHAPTER 8

1. US Geological Survey, "The Cataclysmic 1991 Eruption of Mount Pinatubo, Philippines," US Geological Survey Fact Sheet 113–97, last modified February 28, 2005, https://pubs.usgs.gov/fs/1997/fs113-97/; Carole Volon, Johan Lavreau, and Alain Bernard, "Monitoring Pinatubo Paroxysmal Eruption Plume of June 1991 Using NOAA and GMS Satellite Images," *Proceedings of SPEI* 2309 (December 23, 1994): 321.

2. US Geological Survey, "Cataclysmic 1991 Eruption"; JoAnna Wendel and M. Kumar, "Pinatubo 25 Years Later: Eight Ways the Eruption Broke Ground," *Eos*, June 9, 2016, https://eos.org/articles/pinatubo-25-years-later-eight-ways-the-eruption-broke-ground; US Geological Survey, "Remembering Mount Pinatubo 25 Years Later," June 13, 2016, www.usgs.gov/news/remembering-mount-pinatubo-25-years-ago-mitigating-crisis; Naomi Klein, *This Changes Everything: Capitalism vs. the Climate* (New York: Simon & Schuster, 2014), 258.

3. US Geological Survey, "Remembering Mount Pinatubo"; US Geological Survey, "Cataclysmic 1991 Eruption."

4. John T. Bowen Jr., Thomas R. Leinbach, and Daniel Mabazza, "Air Cargo Services, the State and Industrialization Strategies in the Philippines: The Redevelopment of Subic Bay," *Regional Studies* 36, no. 5 (2002): 456; Karl Schoenberger, "Subic Bay: The Rebirth of a Relic," *Los Angeles Times*, May 5, 1994, WVA1. Karl Schoenberger, staff writer for the *Los Angeles Times*, said a company public relations official told him that Smith's visit to Subic Bay during the Vietnam War did not directly influence his decision to consider doing business there.

5. Bowen, Leinbach, and Mabazza, "Air Cargo Services," 456.

6. David E. Sanger, "Philippines Orders U.S. to Leave Strategic Navy Base at Subic Bay," *New York Times*, December 28, 1991, 1.

7. Bob Drogin, "Americans Bid Farewell to Last Philippine Base," *Los Angeles Times*, November 25, 1992, 232; "'I Shall Return' Deemed Official Version of Vow," *New York Times*, April 6, 1964, 25.

8. Bowen, Leinbach, and Mabazza, "Air Cargo Services," 454–55. For a broader discussion of what some have called "disaster capitalism," see Naomi Klein, *The Shock Doctrine: The Rise of Disaster Capitalism* (New York: Picador, 2008).

9. "Focus: Freight and Transportation—Hubs: Opportunity Knocks," *Far Eastern Economic Review*, March 9, 1995, 48; "Federal Express, Philippines Sign Agreement for Subic Bay," Agence France-Presse, November 13, 1994, n.p.

10. *Subic Bay: The Freeport*, brochure, Subic Bay Metropolitan Authority, 1, 6, copy made available to author by Professor John T. Bowen, Central Washington University; Bowen, Leinbach, and Mabazza, "Air Cargo Services," 457.

11. Gordon quoted in Bowen, Leinbach, and Mabazza, "Air Cargo Services," 457–58.

12. "Evergreen Sells Recently Acquired China Rights to FedEx," *Air Cargo Report* 2, no. 4 (March 2, 1995): n.p.; Rick Brooks, "UPS Seeks to Chip Away at FedEx's Big Lead in China," *Wall Street Journal*, May 16, 2000, B4.

13. House Committee on Transportation and Infrastructure, *Oversight Visit to Wichita, KS, Denver, CO, Seattle, WA, Long Beach, CA, and Memphis, TN, February 18–21, 1997*, 105th Cong., 1st sess., April 1997, 15.

14. "Federal Express's Lobbyists, Led by Chairman, Are Proving to Be Major Force in Washington," *Wall Street Journal*, August 8, 1995, A14.

15. Bill Clinton, speech at Delta Vision, Delta Voices conference, Arlington, Va., May 10, 2000; draft of annotated agenda for Delta Vision, Delta Voices conference, May 10–11, 2000, William J. Clinton Digital Presidential Library. Smith was also an invited guest at the 1995 Southern Regional Economic Conference, a meeting that Vice President Al Gore attended, and Smith played an active role as a panel participant. Memorandum from Jim Kohlenberger for Vice President Al Gore, Re: Regional Economic Conference, March 26, 1995, William J. Clinton Digital Presidential Library.

16. Bill Clinton, speech at World War II memorial groundbreaking, National Mall, Washington, D.C., November 11, 2000, William J. Clinton Digital Presidential Library.

17. Bob Woodward and Ann Devroy, "An Unusual Meeting of Chief Executives," *Washington Post*, August 21, 1997, A1; Asra Q. Nomani and Douglas A. Blackmon, "How Maneuvering by Airlines Shaped U.S.-Japan Accord," *Wall Street Journal*, February 2, 1998, A1.

18. Nomani and Blackmon, "How Maneuvering by Airlines."

19. Woodward and Devroy, "Unusual Meeting."

20. Woodward and Devroy, "Unusual Meeting"; Nomani and Blackmon, "How Maneuvering by Airlines."

21. David E. Sanger, "U.S. Imposes Sanctions against Japan," *New York Times*, September 5, 1997, D1, D4; Mike Esterl, "U.S., Japan Near Accord on Open Skies," *Wall Street Journal*, September 19, 1997, A3.

22. House Committee on Transportation and Infrastructure, Subcommittee on Aviation, *International Aviation Materials and Code Sharing Relationships (Focusing on Japan)*, 105th Cong., 1st sess., June 12, 1997, 39–40, 56.

23. Neil A. Lewis, "A Lobby Effort That Delivers the Big Votes," *New York Times*, October 12, 1996, 37; "Federal Express's Lobbyists."

24. Bipartisan Campaign Reform Act of 1997, Cloture Motion, 105th Cong., 1st sess., October 9, 1997; Lewis, "Lobby Effort That Delivers."

25. Nomani and Blackmon, "How Maneuvering by Airlines"; Matthew L. Wald, "Pact on Air Traffic by U.S. and Japan to Lift Restrictions," *New York Times*, January 31, 1998, A1.

26. Larry Lee Pressler, "Federal Express Hub at Subic Bay," speech, September 12, 1995, 104th Cong., 1st sess., *Congressional Record* 141, pt. 17: 24606; Bowen, Leinbach, and Mabazza, "Air Cargo Services," 459.

27. Robert Heibeler, Thomas B. Kelly, and Charles Ketterman, *Best Practices: Building Your Business with Customer-Focused Solutions* (New York: Touchstone, 1998), 48–49.

28. Draft Electronic Commerce Working Group Report to the President, November 20, 1998, William J. Clinton Digital Presidential Library.

29. "FedEx Customer Success Story: Philips Semiconductor," FedEx, 2002, accessed October 4, 2022, www.fedex.com/us/supplychain/casestudies/philips.pdf; "Warehouses That Fly," *Forbes*, October 18, 1999, www.forbes.com.

30. R. Chepesiuk, "Where the Chips Fall: Environmental Health in the Semiconductor Industry," *Environmental Health Perspectives* 107, no. 9 (September 1999): A456; Yun-Sung Eom et al., "Emission Factors of Air Toxics from Semiconductor Manufacturing in Korea," *Journal of the Air and Waste Management Association* 56 (2006): 1518–24.

31. Zoë Schlanger, "Silicon Valley Is Home to More Toxic Superfund Sites than Anywhere Else in the Country," *Quartz*, June 28, 2017, https://qz.com/1017181/silicon-valley-pollution-there-are-more-superfund-sites-in-santa-clara-than-any-other-us-county/.

32. Cam Simpson, "American Chipmakers Had a Toxic Problem. Then They Outsourced It," *Bloomberg Businessweek*, June 15, 2017, www.bloomberg.com/news/features/2017-06-15/american-chipmakers-had-a-toxic-problem-so-they-outsourced-it.

33. Simpson, "American Chipmakers."

34. "8 Things You Should Know about Water and Semiconductors," China Water Risk, July 11, 2013, http://chinawaterrisk.org/resources/analysis-reviews/8-things-you-should-know-about-water-and-semiconductors/.

35. Seth Mydans, "Subic Bay, Minus U.S., Becomes Surprise Success," *New York Times*, November 23, 1996, 3; Claire Wallerstein, "Subic Bay Thrives in Post-US Era," *Christian Science Monitor*, November 10, 1997, https://www.csmonitor.com/1997/1110/111097.intl.intl.2.html.

36. On Japan Airlines and other early pioneers in the fish flying business, see Sasha Isenberg, *The Sushi Economy: Globalization and the Making of a Modern Delicacy* (New York: Gotham Books, 2007). I am deeply indebted to former FedEx station manager and PeriShip CEO Luciano Morra for talking with me about fish and perishable food shipments. After leaving FedEx in the late 1990s, Morra started PeriShip, working exclusively with FedEx to help companies ship perishable goods. Author interview with Luciano Morra, July 25, 2018. Business historian Manse Blackford discusses PeriShip and FedEx in his book on the fishing industry, *Making Seafood Sustainable: American Experiences in Global Markets* (Philadelphia: University of Pennsylvania Press, 2012), 179. On decline of bluefin tuna and other Pacific fish populations, see Lucy Siegle, "Can Sushi Ever Be Sustainable?," *The Guardian*, May 17, 2015, www.theguardian.com/environment/2015/may/17/is-sushi-sustainable.

37. "UPS: New Planes Raise Safety Issues," *Atlanta Constitution*, November 29, 2000, D19; Bowen, Leinbach, and Mabazza, "Air Cargo Services," 465.

38. Bowen, Leinbach, and Mabazza, "Air Cargo Services," 465, 461.

39. FedEx, "FedEx Initiates Flight Operations Test at Its New Asia Pacific Hub in Baiyun International Airport in Guangzhou" (press release), December 17, 2008, https://newsroom.fedex.com/newsroom/fedex-initiates-flight-operations-test-at-its-new-asia-pacific-hub-in-baiyun-international-airport-in-guangzhou/.

40. Author interview with Richard Smith, regional president of the Americas and executive vice president of FedEx, March 15, 2021.

41. FedEx, "FedEx Helping a Pharmaceutical Customer Deliver Globally," YouTube video, 3:37, posted by FedEx APAC on May 9, 2018, https://www.youtube.com/watch?v=_s4a5WhJAD8.

42. FedEx, *Thinking outside the Box: 2012 Report on Global Programs in Disaster Readiness, Relief and Recovery*, June 2012, 11, document in author possession.

43. Senate Committee on Homeland Security and Governmental Affairs, Permanent Subcommittee on Investigations, *Combatting the Opioid Crisis: Exploiting Vulnerabilities in International Mail*, 115th Cong., 2nd sess., January 25, 2018, 6, 61, 73, 78, 204.

44. Simon Denyer, "Even as China Turns Away from Shark Fin Soup, the Prestige Dish Is Gaining Popularity Elsewhere in Asia," *Washington Post*, February 15, 2018, www.washingtonpost.com; "The Buck Stops with Shark Fin Supply Chain," video, Reuters, July 26, 2016, https://web.archive.org/web/20160710092018/https://www.reuters.com/video/2016/07/07/the-buck-stops-with-shark-fin-supply-cha?videoId=369197951&videoChannel=102&channelName=Entertainment.

45. Denyer, "Shark Fin Soup"; "Buck Stops."

46. Ernest Kao, "FedEx under Pressure to Follow Carriers and Ban Shark Fin Cargo as Activists Picket Hong Kong Depot," *South China Morning Post*, August 13, 2018, www.scmp.com/news/hong-kong/health-environment/article/1984800/fedex-under-pressure-follow-carriers-and-ban-shark.

47. "Buck Stops."

48. 2017 FedEx Global Citizenship Report, 18, 28–45, https://www.fedex.com/content/dam/fedex/us-united-states/sustainability/gcrs/FedEx_2017_Global_Citizenship_Report.pdf.

49. 2022 FedEx ESG Report, 5, 15, 20, 36–37; 2021 FedEx ESG Report, 34; 2018 FedEx ESG Report, 58–59; 2017 FedEx Global Citizenship Report, 33; all available on the FedEx website under "Environmental, Social, and Governance Reporting," https://www.fedex.com/en-us/sustainability/reports.html.

50. House Committee on Energy and Commerce, Subcommittee on Oversight and Investigations, *Safety of Imported Pharmaceuticals: Strengthening Efforts to Combat the Sales of Controlled Substances over the Internet*, 109th Cong., 1st sess., December 13, 2005, 182. In 2016, NPR reported that the Justice Department had issued an indictment in 2014 against FedEx for "allegedly shipping packages from illegal internet pharmacies" but noted that those charges were later dropped. Rebecca Hersher, "Justice Department Drops Charge That FedEx Shipped for Illegal Pharmacies," *The Two-Way*, NPR, June 17, 2016, https://www.npr.org/sections/thetwo-way/2016/06/17/482537913/justice-department-drops-charge-that-fedex-shipped-for-illegal-pharmacies.

CHAPTER 9

1. Author interview with Hugh McColl, July 26, 2019, Bank of America Corporate Center, Charlotte, N.C. I would like to offer special thanks to the editors of *Environmental History*, Mark Hersey and Stephen Brain, who kindly allowed me to reprint sections of an article in this chapter and the next. For that original article, see Bartow J. Elmore, "The Environmental History of an American Bank," *Environmental History* 27, no. 1 (January 2022): 113–39.

2. Author interview with Hugh McColl.

3. Author interview with Hugh McColl.

4. Thomas I. Storrs, "Profitability and Responsibility," speech at graduation ceremonies for Stonier Graduate School of Banking, Rutgers University, New Brunswick, N.J., June 23, 1972, box 1, folder: Storrs Speeches, 1971–1972, NationsBank Records.

5. The environmental history of banking has remained a remarkably unexplored area of research among historians. Over the past two decades, scholars exploring the nexus of business and environmental history have produced numerous works that have examined the relationships between corporations and the ecosystems they inhabit. For early works calling for a merger of business and environmental history scholarship, see Christine Meisner Rosen and Christopher Sellers, "The Nature of the Firm: Towards an Ecocultural History of Business," *Business History Review* 73 (Winter 1999): 577–600; and Christine Meisner Rosen, "The Business-Environment Connection," *Environmental History* 10 (2005): 77–79. In 2017, Hartmut Berghoff and Adam Rome edited

Green Capitalism? Business and the Environment in the Twentieth Century (Philadelphia: University of Pennsylvania Press, 2017). This excellent volume offers new articles on business and environmental history as well as succinct digestions of recent scholarship that draws on these two subfields. Notably, *Green Capitalism?* does not include any significant discussion of the environmental history of the financial industry. Most of this scholarship has focused on businesses directly engaged in natural resource extraction and processing. We now know a great deal about the ecological footprints of firms that pumped oil from the ground, transported bananas from tropical lands, and produced the chemicals that made our modern world. See, for example, John Soluri, *Banana Cultures: Agriculture, Consumption and Environmental Change in Honduras and the United States* (Austin: University of Texas Press, 2006); Richard Tucker, *Insatiable Appetite: The United States and the Ecological Degradation of the Tropical World* (Berkeley: University of California Press, 2000); and Gregg Mittman, *Empire of Rubber: Firestone's Scramble for Land and Power in Liberia* (New York: New Press, 2021). Notable works looking at industrial pollution include Christopher Sellers, *Hazards of the Job: From Industrial Disease to Environmental Health Science* (Chapel Hill: University of North Carolina Press, 1999); Gerald Markowitz and David Rosner, *Deceit and Denial: The Deadly Politics of Industrial Pollution* (Berkeley: University of California Press, 2002); Hugh S. Gorman, *Redefining Efficiency: Pollution Concerns, Regulatory Mechanisms, and Technological Change in the U.S. Petroleum Industry* (Akron: University of Akron Press, 2001); and Ellen Spears, *Baptized in PCBs: Race, Pollution and Justice in an American Town* (Chapel Hill: University of North Carolina Press, 2014). See also Adam Rome, "DuPont and the Limits of Corporate Environmentalism," *Business History Review* 93 (Spring 2019): 75–99. Scholars have also provided a clearer picture of the history of environmental corporate social responsibility programs that emerged in the 1980s and 1990s, thanks to works such as Archie B. Carroll, Kenneth J. Lipartito, James E. Post, and Patricia H. Werhane, *Corporate Responsibility: The American Experience* (New York: Cambridge University Press, 2012); and Adam Rome, "Beyond Compliance: The Origins of Corporate Interest in Sustainability," *Enterprise and Society* 22 (June 2021): 409–37. Yet this new scholarship deals very little with commercial banks or other investment firms, which is surprising because some of the most radical ecological transformations of the past few decades would not have been possible without the megabanks that channeled tremendous capital toward Earth-changing undertakings. Books about the "greening" of global banking abound, but most of these studies have been written by business and finance professionals or business school scholars and include only limited historical analysis. See, for example, Stephan Schmidheiny and Federico Zorraquín, *Financing Change: The Financial Community, Eco-Efficiency, and Sustainable Development* (Cambridge, Mass.: MIT Press, 1996); Olaf Weber and Blair Feltmate, eds., *Sustainable Banking: Managing the Social and Environmental Impact of Financial Institutions* (Toronto: University of Toronto Press, 2016); and Marcel Jeucken, *Sustainable Finance and Banking: The Finance Sector and the Future of the Planet* (London: Earthscan, 2010). A notable exception would be the work of the business historian Geoffrey Jones. See, for example, Geoffrey Jones, "Can Finance Change the World?," in *Profits and Sustainability: A History of Green Entrepreneurship* (New York: Oxford University Press, 2017), 270–309; and Geoffrey Jones, Emily Grandjean, and Andrew Spadafora, "Financing Sustainability," in *Varieties of Green Business: Industries, Nations, and Time* (Northampton, Mass.: Edward

Elgar, 2018), 93–121. In short, business and environmental historians have said very little about the ecological footprint of large megabanks and even less about environmental activism that targeted these firms in recent decades.

6. Memorandum from D. D. McColl to Stockholders of the Bank of Marlboro, date unknown, Subseries 2.1: Bank of Marlboro and Related Material, 1866–1990 and undated, box 8, folder: 1866–1893, McColl Papers; transcription of Hugh McColl interview conducted by Joe Martin, 1996, box 3, folder: McColl, Hugh circa 1996, NationsBank Records. For a nice summation of D. D. McColl's life, see Ross Yockey, *McColl: The Man with America's Money* (Atlanta: Longstreet, 1999), 76–85.

7. Yockey, *McColl*, 81; memorandum to members of the Board of Trustees, January 2, 1917, Subseries 2.1: Bank of Marlboro and Related Material, 1866–1990 and undated, box 8, folder: 1917–1919, McColl Papers.

8. Yockey, *McColl*, 81–82, 85; transcription of Hugh McColl interview conducted by Joe Martin.

9. Howard E. Covington Jr. and Marion A. Ellis, *The Story of NationsBank: Changing the Face of American Banking* (Chapel Hill: University of North Carolina Press, 1993), 27; "NCNB, a Brief History," company publication issued in 1977, box 2, folder: A Brief History (1977), NationsBank Records.

10. Covington and Ellis, *Story of NationsBank*, 27; Doug Campbell, "Branch by Branch: How North Carolina Became a Banking Giant," *Region Focus* 10 [a publication of the Federal Reserve Bank of Richmond] (Fall 2006): 53; Thomas D. Hills, "The Rise of Southern Banking and the Disparities among the States following the Southeastern Regional Company," *North Carolina Banking Institute* 11 (2007): 95. I would like to thank Thomas Hills for discussing the history of North Carolina banking in an interview with me on March 10, 2020. I am also indebted to University of North Carolina law professor Lissa Broome, who discussed the history of branch banking in an author interview conducted on July 12, 2019.

11. Campbell, "Branch by Branch," 53.

12. Transcription of interview with Tom Storrs, September 17, 1996, box 3, folder: Storrs, Tom 17 September, 1996, NationsBank Records. On American Trust's correspondent banking strategy, see Covington and Ellis, *Story of NationsBank*, 26–27.

13. Letter from W. H. Wood, president of American Trust Company, to Hugh McColl Jr., January 10, 1931, Subseries 2.1: Bank of Marlboro and Related Material, 1866–1990 and undated, box 9, folder: 89, McColl Papers; Yockey, *McColl*, 17, 86. For an excellent environmental history of the boll weevil's spread through the American South, see Jim Giesen, *Boll Weevil Blues: Cotton, Myth, and Power in the American South* (Chicago: University of Chicago Press, 2011), which is referenced in the first chapter on Delta Air Lines.

14. Letter from David K. McColl to Pearl M. McLeod, December 11, 1933, Subseries 2.1: Bank of Marlboro and Related Material, 1866–1990 and undated, box 9, folder: 97, McColl Papers; Yockey, *McColl*, 93.

15. Hugh L. McColl obituary, *Greenville (S.C.) News*, July 5, 1994, 5C; transcription of Hugh McColl interview conducted by Joe Martin; Yockey, *McColl*, 92–93. For discussion of the Marlboro Mills strike, see Anthony P. Dunbar, *Against the Grain: Southern Radicals and Prophets, 1929–1959* (Charlottesville: University Press of Virginia, 1981), 145–46.

16. Gerald D. Nash, *A. P. Giannini and the Bank of America* (Norman: University of Oklahoma Press, 1992), 6–8; Marquis James and Bessie Rowland James, *A Biography of a Bank: The Story of Bank of America N. T. & S. A.* (New York: Harper & Brothers, 1954), 4–5. On refrigerated railcars see Ted Steinberg, *Down to Earth: Nature's Role in American History*, 3rd ed. (New York: Oxford University Press, 2013), 175.

17. Nash, *A. P. Giannini*, 8–9; James and James, *Biography of a Bank*, 5.

18. James and James, *Biography of a Bank*, 5–6; Nash, *A. P. Giannini*, 9–12, 17.

19. Nash, *A. P. Giannini*, 14, 16–17; James and James, *Biography of a Bank*, 9.

20. Nash, *A. P. Giannini*, 12–15, 17–18. On the history of irrigation projects in California and the American West at this time, see Donald Worster, *Rivers of Empires: Water, Aridity, and the Growth of the American West* (New York: Oxford, 1992); and Donald J. Pisani, *From Family Farm to Agribusiness: The Irrigation Crusade in California and the West, 1850–1931* (1984; repr., Berkeley: University of California Press, 2021).

21. James and James, *Biography of a Bank*, 9–15; Nash, *A. P. Giannini*, 23–25.

22. Nash, *A. P. Giannini*, 23–25. On the history of loan sharks and the state of consumer credit markets in the Gilded Age and Progressive Era, see Louis Hyman, *Debtor Nation: The History of America in Red Ink* (Princeton, N.J.: Princeton University Press, 2011), 10–44.

23. James and James, *Biography of a Bank*, 13–15, 204, 212; Nash, *A. P. Giannini*, 97.

24. Nash, *A. P. Giannini*, 29–30.

25. "Biggest Bank: Bank of America Rolls Up Its Sleeves to Keep Lead over Rivals in East," *Wall Street Journal*, April 15, 1947, 1, 4; Nash, *A. P. Giannini*, 5, 30–35.

26. Nash, *A. P. Giannini*, 32.

27. On Giannini's response after the San Francisco earthquake, see James and James, *Biography of a Bank*, 25–28; Nash, *A. P. Giannini*, 5, 30–35; Yockey, *McColl*, 554.

28. Nash, *A. P. Giannini*, 39.

29. Nash, *A. P. Giannini*, 37–38; James and James, *Biography of a Bank*, 40–41.

30. James and James, *Biography of a Bank*, 52; Nash, *A. P. Giannini*, 42.

31. Nash, *A. P. Giannini*, 93; Judge Earl Glock, *The Dead Pledge: The Origins of the Mortgage Market and Federal Bailouts* (New York: Columbia University Press, 2021), 4, 9, 60–85. For more on the Federal Land Banks that were set up by the federal government after 1916, see Glock, *Dead Pledge*, 12–31.

32. Nash, *A. P. Giannini*, 45–47; James and James, *Biography of a Bank*, 52, 72, 80.

33. James and James, *Biography of a Bank*, 93, 113.

34. James and James, *Biography of a Bank*, 110, 259, 262–63.

35. Moira Johnston, *Roller Coaster: The Bank of America and the Future of American Banking* (New York: Ticknor & Fields, 1990), 41; Yockey, *McColl*, 554; James and James, *Biography of a Bank*, 167–68.

36. Johnston, *Roller Coaster*, 41–42; James and James, *Biography of a Bank*, 106, 202; 1930 Bank of America Annual Report, 2.

37. "Biggest Bank"; 1940 Bank of America Annual Report, 10–11; Nash, *A. P. Giannini*, 114–18, 136; James and James, *Biography of a Bank*, 417, 490.

38. Adam Rome, "Building on the Land: Toward an Environmental History of Residential Development in American Cities and Suburbs, 1870–1990," *Journal of Urban History* 20, no. 3 (May 1994): 416.

39. "Biggest Bank"; Nash, *A. P. Giannini*, 114–18, 136; James and James, *Biography of a Bank*, 418–19, 490; Johnston, *Roller Coaster*, 22.

40. Nash, *A. P. Giannini*, 144; Johnston, *Roller Coaster*, 21; 1949 Bank of America Annual Report, 1.

41. Hills, "Rise of Southern Banking," 61.

42. Transcription of interview with Tom Storrs, September 17, 1996, box 3, folder: Storrs, Tom, 17 September 1996, NationsBank Records; Covington and Ellis, *Story of NationsBank*, 61.

43. "A Brief History," 1977, box 2, folder: A Brief History, 1977, NationsBank Records; interview with George Snyder, box 2, folder: Transcript of Interview—George Snyder, NationsBank Records; Yockey, *McColl*, 17; Rick Rothacker, *Banktown: The Rise and Struggles of Charlotte's Big Banks* (Winston-Salem, N.C.: John F. Blair, 2010), 11.

44. Interview with Hugh McColl, 1983, box 2, folder: Transcript of Interview—Hugh McColl, 1983, NationsBank Records; Yockey, *McColl*, 30–31; Covington and Ellis, *Story of NationsBank*, 56.

45. Letter from Hugh McColl to father, January 21, 1955, Subseries 1.1: Family Correspondence, 1834–1995 and undated (RESTRICTED), box 3, folder: 1954–1957, McColl Papers.

46. Yockey, *McColl*, 36, 39–40.

47. Yockey, *McColl*, 39–40, 62; author interview with Hugh McColl.

48. Author interview with Hugh McColl.

49. Covington and Ellis, *Story of NationsBank*, 24–26.

50. C. M. Vanstory, president of Security National, admitted in an interview that Charlotte being in a Federal Reserve city was the reason Security National came to the negotiating table. Interview with C. M. Vanstory, box 2, folder: Transcript of Interview with C. M. Vanstory, 1983, NationsBank Records; interview with Jon Van Lindley, 1983, box 2, folder: Transcript of Interview—John Van Lindley, 1983, NationsBank Records; Rothacker, *Banktown*, 4; Covington and Ellis, *Story of NationsBank*, 40.

51. Yockey, *McColl*, 33.

52. Author interview with Hugh McColl; For more on the political climate in Charlotte in the 1960s and the city's move to desegregate public accommodations and schools, see Matthew D. Lassister, *The Silent Majority: Suburban Politics in the Sunbelt South* (Princeton, N.J.: Princeton University Press, 2006), 121–47.

53. Confidential memorandum from Thomas Storrs to Mr. Dougherty et al., October 25, 1976, box 1, folder: Storrs Correspondence, 1976, NationsBank Records.

54. Storrs, "Profitability and Responsibility."

55. Edward L. Ayers, *Southern Journey: The Migrations of the American South, 1790–2020* (Baton Rouge: Louisiana State University Press, 2020), 93–94, 113.

56. Author interview with Hugh McColl.

57. "NCNB, a Brief History," company publication issued in 1977, box 2, folder: A Brief History (1977), NationsBank Records.

58. For an excellent history of Bank of America's credit card launch in Fresno, see "The Drop" in Joe Nocera's *A Piece of the Action: How the Middle Class Joined the Money Class* (New York: Simon and Schuster, 2013), 15–33.

59. Thomas I. Storrs, "Innovate or Liquidate," speech, Bank Administration Institute conference, Fredericksburg, Va., May 1, 1971, box 1, folder: Storrs Speeches, 1971–72, NationsBank Records; Covington and Ellis, *Story of NationsBank*, 76.

60. Thomas Storrs, "Responsibilities of Private Enterprise for Growth in the Piedmont Crescent," speech, n.d., box 1, folder: Storrs Speeches, 1962–1968,

NationsBank Records; Covington and Ellis, *Story of NationsBank*, 63; Yockey, *McColl*, 63, 73, 173.

61. Storrs, "Responsibilities of Private Enterprise."

62. Storrs, "Profitability and Responsibility."

63. Letter from Thomas I. Storrs to Subsidiary Presidents, Division Executives, Regional Executives, City Executives, November 30, 1973, box 1, folder: Storrs Correspondence, 1973, NationsBank Records; letter from Thomas I. Storrs to Council on Environmental Quality, Washington, D.C., October 3, 1973, box 1, folder: Storrs Correspondence, 1973, NationsBank Records.

64. Johnston, *Roller Coaster*, 86; Walter Russell Mead, "The Bank of America's Clausen: When a Golden Boy Lost His Gleam," *Los Angeles Times*, October 4, 1987, E1; James and James, *Biography of a Bank*, 17; Nash, *A. P. Giannini*, 141.

65. Mead, "Bank of America's Clausen"; "Bank of America, 14 Others Plan Loan to Bolivia's Oil Firm," *Wall Street Journal*, July 1, 1975, 22; Johnston, *Roller Coaster*, 87–91, 175; "Bank of America Announces Loan to Mexican Oil Firm," *Wall Street Journal*, April 26, 1968, 26; "Bank of America Boosts Credit Line to Petrobas," *Wall Street Journal*, January 16, 1976, 2; "Bank of America Group Provides Loans to Brazil," *Wall Street Journal*, January 12, 1981, 30; "Bank of America to Advise Alaska," *New York Times*, August 13, 1969, 59.

66. Johnston, *Roller Coaster*, 130; Covington and Ellis, *Story of NationsBank*, 20, 273; Dermot Gately, "Lessons from the 1986 Oil Price Collapse," *Brookings Institute Papers on Economic Activity* 2 (1986): 250, www.brookings.edu/wp-content/uploads/1986/06/1986b_bpea_gately_adelman_griffin.pdf; Yockey, *McColl*, 315–16.

67. Interview with Hugh McColl, September 15, 1997, box 3, folder: September 15, 1997, NationsBank Records; interview with Joe Martin, July 15, 1983, box 2, folder: Transcript of Interview—Joe Martin, NationsBank; Thomas I. Storrs, "Interstate Banking—It Has Arrived," January 5, 1984, box 1, folder: Storrs Correspondence, 1984, NationsBank Records; letter from Thomas I. Storrs to J. Joseph Tuchy, chairman of Landmark First National Bank, December 21, 1981, box 1, folder: Storrs Correspondence, 1981, NationsBank Records; "NCNB Corp. Is Cleared by Fed to Buy Control of Lake City, Fla., Bank," *Wall Street Journal*, December 10, 1981, 6; Hills, "Rise of Southern Banking," 57–104; Jeremy Markovich, "The Man Who Built Charlotte," *Our State*, October 23, 2017, www.ourstate.com/hugh-mccoll-charlotte-banking/; Yockey, *McColl*, 300.

68. Rothacker, *Banktown*, 20; Covington and Ellis, *Story of NationsBank*, 213, 314.

69. On the inner workings of the FDIC deal and tax breaks, see interview with Hugh McColl, September 15, 1997; Rothacker, *Banktown*, 20; Yockey, *McColl*, 320–21, 330, 340, 344; Hills, "Rise of Southern Banking," 80, 86; Covington and Ellis, *Story of NationsBank*, 213–14.

70. Hills, "Rise of Southern Banking," 86; Covington and Ellis, *Story of NationsBank*, 255; Yockey, *McColl*, 356.

71. James Bates, "Lenders Balk at Pumping More Money into Oil," *Los Angeles Times*, August 28, 1990, D2; Hills, "Rise of Southern Banking," 57; Covington and Ellis, *Story of NationsBank*, 2.

72. Campbell, "Branch by Branch," 53; Yockey, *McColl*, 504, 508; Saul Hansell, "NationsBank to Acquire Bank South," *New York Times*, September 6, 1995, D1; 1996 NationsBank Annual Report, 13. On the relationship with President Clinton, see Yockey, *McColl*, 470–75, 484; and Rothacker, *Banktown*, 23.

CHAPTER 10

1. Walter Russell Mead, "The Bank of America's Clausen: When a Golden Boy Lost His Gleam," *Los Angeles Times*, October 4, 1987, E1; FDIC, Division of Research and Statistics, *Histories of the Eighties—Lessons for the Future*, vol. 1, *An Examination of the Banking Crises of the 1980s and Early 1990s*, December 1997, 191, www.fdic.gov/bank/historical/history/vol1.html; Moira Johnston, *Roller Coaster: The Bank of America and the Future of American Banking* (New York: Ticknor & Fields, 1990), 349.

2. "Banking: Calm Follows a Crisis of Confidence," *Los Angeles Times*, January 27, 1985, C1; Johnston, *Roller Coaster*, 302; 1986 Bank of America Annual Report, 1; Mark Potts, "Bank of America Struggles to Regain Financial Stability," *Washington Post*, May 17, 1987, 157; Mead, "Bank of America's Clausen"; Pamela Riney-Kehrberg, "Children of the Crisis: Farm Youth in Troubled Times," *Middle West Review* 2, no. 1 (Fall 2015): 11–25; Gilbert C. Fite, "The Farm Debt Crisis of the 1980s: A Review Essay," *Annals of Iowa* 51, no. 3 (Winter 1992): 288–93.

3. Mead, "Bank of America's Clausen"; Nancy Rivera Brooks, "Farmers Finding Hard Row to Hoe," *Los Angeles Times*, July 12, 1991, D1; Maria L. La Ganga, "Farmers Also Face a Loan Drought," *Los Angeles Times*, March 28, 1991, D1.

4. Adam Rome, "Beyond Compliance: The Origins of Corporate Interest in Sustainability," *Enterprise and Society* 22 (June 2021): 413–15; "Exxon Valdez," Damage Assessment, Remediation, and Restoration Program, National Oceanic and Atmospheric Administration, last updated August 17, 2020, https://darrp.noaa.gov/oil-spills/exxon-valdez; Alan Taylor, "Bhopal: The World's Worst Industrial Disaster, 30 Years Later," *The Atlantic*, December 2, 2014, www.theatlantic.com/photo/2014/12/bhopal-the-worlds-worst-industrial-disaster-30-years-later/100864/; "Grief Turns to Anxiety in Bhopal," *Wall Street Journal*, January 27, 1985, A1.

5. Bank of America want ad for environmental analyst, *Los Angeles Times*, April 2, 1989, 62; Bank of America want ad for environmental analyst, *Los Angeles Times*, July 30, 1991, 8; "Toxic Liability May Dampen Farmers' Ability to Get Loans," *Lompoc (Calif.) Recorder*, July 31, 1990, A6; Richard N. L. Andrews, *Managing the Environment, Managing Ourselves* (1999; repr., New Haven, Conn.: Yale University Press, 2008), 248–49, 265; Bank of America, "Bank of America Sets New Industry Best Practices for Climate Change and Forest Policies" (press release), PR Newswire, May 27, 2004, 1; Tom Zeller Jr., "Lenders Step Away from Environmental Risks," *New York Times*, August 31, 2010, A1; "Ecological Fumbles Soil Bottom Line," *San Francisco Examiner*, February 16, 1997, W-34; Matthew H. Ahrens and David S. Langer, "Liability under CERCLA, Environmental Risks for Lenders under Superfund: A Refresher for the Economic Downturn," *Bloomberg Corporate Law Journal* 3 (2008): 482–93; Geoffrey Jones, *Profits and Sustainability: A History of Green Entrepreneurship* (New York: Oxford University Press, 2017), 272.

6. Jones, *Profits and Sustainability*, 274.

7. "The Business and Politics of Energy," *Buildings* 92, no. 8 (August 1998): 84–86.

8. James Bates, "Banking on a Grand Scale," *Los Angeles Times*, October 18, 1992, SM12.

9. Matthew E. Kahn, "The Environmental Impact of Suburbanization," *Journal of Policy Analysis and Management* 19, no. 4 (Autumn 2000): 570.

10. Adam Rome, "Building on the Land: Toward an Environmental History of Residential Development in American Cities and Suburbs, 1870–1990," *Journal of Urban History* 20, no. 3 (May 1994): 419. For more on the environmental effects of suburbanization, see Adam Rome, *Bulldozer in the Countryside: Suburban Sprawl and the Rise of Environmentalism* (New York: Cambridge University Press, 2001); Christopher Sellers, *Crabgrass Crucible: Suburban Nature and the Rise of Environmentalism in Twentieth-Century America* (Chapel Hill: University of North Carolina Press, 2012); and Andrew C. Baker, *Bulldozer Revolutions: A Rural History of the Metropolitan South* (Athens: University of Georgia Press, 2018).

11. Frank Clifford, "Sprawl's Costs Hurting State, Report Finds," *Los Angeles Times*, January 31, 1995, A3.

12. 2007 Bank of America Annual Report, 32; Jones, *Profits and Sustainability*, 272; author interview with Michael Brune, executive director of the Sierra Club, July 17, 2021. For more on the UNEP FI, see www.unepfi.org/about/.

13. Michelle Singletary, "Bank of America Ventures East to Open D.C. Office," *Washington Post*, December 25, 1995, 10; Chris Kraul, "NationsBank and BofA: Is Bigger Really Better?," *Los Angeles Times*, October 24, 1995, D1; Rick Brooks, "NationsBank, BankAmerica Holders Vote for $43.01 Billion Merger Plan," *Wall Street Journal*, September 25, 1998, B6.

14. Kraul, "NationsBank and BofA"; Brooks, "NationsBank, BankAmerica Holders Vote"; 1999 Bank of America Annual Report, 29; author interview with Hugh McColl, July 26, 2019, Bank of America Corporate Center, Charlotte, N.C.; Rick Rothacker, *Banktown: The Rise and Struggles of Charlotte's Big Banks* (Winston-Salem, N.C.: John F. Blair, 2010), 23, 26; Yockey, *McColl*, 555–60; 1998 Bank of America Annual Report, 4.

15. Fortune 500 list for 1980, accessed October 24, 2022, https://archive.fortune.com/magazines/fortune/fortune500_archive/full/1980/; Fortune 500 list for 2010, accessed October 24, 2022, https://money.cnn.com/magazines/fortune/fortune500/2010/full_list/.

16. 2010 Rainforest Action Network Annual Report, 4, 8; author interview with Michael Brune, former executive director of RAN, July 15, 2021.

17. Yaroslav Trofimov and Helene Cooper, "Antiglobalization Activists Are Shifting Focus to Multinational Corporations," *Wall Street Journal*, July 23, 2001, www.iatp.org/news/antiglobalization-activists-are-shifting-focus-to-multinational-corporations; author interview with Michael Brune, former executive director of RAN, July 15, 2021.

18. Steve Chappelle, "Laws of the Jungle," *Los Angeles Times Magazine*, August 8, 2004, 8–10, 30; 2010 Rainforest Action Network Annual Report, 8.

19. Hiroko Tabuchi, "The Banks Putting Rainforests in Peril," *New York Times*, December 4, 2016, BU1; Bank of America, "Bank of America Sets New Industry Best Practices for Climate Change and Forest Policies" (press release), PR Newswire, May 17, 2004.

20. "BofA Announces Emissions, Rainforest Goals," *Charlotte Observer*, May 18, 2004, 6D.

21. Bank of America, "Bank of America Sets New Industry"; RAN display ad in *New York Times*, May 21, 2004, A11; Jones, *Profits and Sustainability*, 290–94; Archie B. Carroll, Kenneth J. Lipartito, James E. Post, and Patricia H. Werhane, *Corporate Responsibility: The American Experience* (New York: Cambridge University Press, 2012), 366. For more on CERES, see www.ceres.org/about-us.

22. Alison Leigh Cowan, "Taking Protest to a Corporate Chief's Street, 3 Activists Face Charges in Greenwich," *New York Times*, March 13, 2005, 40; "Environmentalists Bypass Washington to Pressure Polluting Corporations," *Red Deer (Alberta, Canada) Advocate*, May 27, 2005, B9.

23. Claudia H. Deutsch, "Gas Emissions Rarely Figure in Investor Decisions," *New York Times*, September 25, 2007, C2.

24. "4 Arrested after Protests of BofA That Closed Uptown Streets," *Charlotte Observer*, October 24, 2007, B2.

25. 2007 Bank of America Annual Report, 32; 2009 Bank of America Annual Report, 20; 2013 Bank of America Corporate Social Responsibility Executive Summary, 4; Bank of America, "Bank of America Launches Brighter Planet Credit Card," (press release), PR Newswire, November 29, 2007; Bank of America display ad in *New York Times*, November 1, 2009, 25.

26. 2010/2011 Rainforest Action Network Annual Report, 3, 13.

27. Rainforest Action Network, BankTrack, and Sierra Club, *Policy and Practice: Report Card on Banks and Mountaintop Removal*," n.p., 2010, www.banktrack.org/download/policy_and_practice_report_card_on_banks_and_mountaintop_removal/100513_report_card_on_banks_and_mtr.pdf. On Sierra Club's Beyond Coal campaign, see Sierra Club, "Sierra Club's Beyond Coal Campaign Marks 350th Coal Plant Set for Retirement," December 15, 2021, https://www.sierraclub.org/articles/2021/12/sierra-club-s-beyond-coal-campaign-marks-350th-coal-plant-set-for-retirement.

28. 2011/2012 Rainforest Action Network Annual Report, 12–13; author interview with Amanda Starbuck, former climate and energy program director at RAN, July 16, 2021.

29. Nelson D. Schwartz, "Banks Look to Burnish Their Images by Backing Green Technology Firms," *New York Times*, June 11, 2012, B1.

30. 2012/2013 Rainforest Action Network Annual Report, 15.

31. 2010 Bank of America Corporate Social Responsibility Report, 10; 2013 Bank of America Corporate Social Responsibility Executive Summary, 4.

32. John Downey, "Duke Energy, BofA Make Dow Jones Sustainability Index," *Charlotte (N.C.) Business Journal*, September 10, 2015, www.bizjournals.com/charlotte/blog/energy/2015/09/duke-energy-bofa-make-dow-jones-sustainability.html; "Bank of America Ranked First among Financial Institutions in Carbon Disclosure Project's 2011 Global 500 and S&P 500 Ranking," *Dow Jones Institutional News*, September 20, 2011; Jones, *Profits and Sustainability*, 288; White House, "Fact Sheet: White House Opens Business Act on Climate Pledge" (press release), July 27, 2015, https://obamawhitehouse.archives.gov/the-press-office/2015/07/27/fact-sheet-white-house-launches-american-business-act-climate-pledge.

33. Rainforest Action Network, *Banking on Climate Change: Fossil Fuel Finance Report Card 2019*, March 20, 2019, 3–4, 7, 45, https://www.ran.org/wp-content/uploads/2019/03/Banking_on_Climate_Change_2019_vFINAL1.pdf; 2016 Bank of America Environmental, Social, and Governance Report, 1.

34. Rainforest Action Network, *Banking on Climate Change: Fossil Fuel Finance Report Card 2020*, March 18, 2020, 54, https://www.ran.org/wp-content/uploads/2020/03/Banking_on_Climate_Change__2020_vF.pdf; 2019 Bank of America Annual Report, 6, 28; 2018 Bank of America Corporation Environmental, Social, and Governance Performance Data Summary, 7.

35. Rainforest Action Network, *Banking on Climate Change: Fossil Fuel Finance Report 2020*; Indigenous Environmental Network, "Banking on Climate Change: Fossil Fuel Report Cards 2018–2020," accessed October 24, 2022, www.ienearth.org/banking-on-climate-change-fossil-fuel-report-cards/.

36. Lennox Yearwood Jr. and Bill McKibben, "Want to Do Something about Climate Change? Follow the Money," *New York Times*, January 11, 2020.

37. Chris Flood, "Big Investors Take Aim at Banks over Climate Change Risk," *Financial Times*, September 14, 2017. For more on ShareAction and the Investor Decarbonization Initiative, see https://shareaction.org/decarbonise/, accessed October 24, 2022. Laurel Wamsley, "World's Largest Asset Manager Puts Climate at Center of Its Investment Strategy," NPR, January 14, 2020, www.npr.org/2020/01/14/796252481/worlds-largest-asset-manager-puts-climate-at-the-center-of-its-investment-strate. On the history of socially responsible investing, which dates to at least the 1970s, see Carroll et al., *Corporate Responsibility*, 394.

38. Wamsley, "World's Largest Asset Manager."

39. Coral Davenport, "Climate Change Poses Risks to Financial System, US Bank CEOs Tell Congress," *SNL Financial Extra*, April 10, 2019.

40. Hannah Lang, "Warren Presses Big-Bank CEOs on Climate Change Policies," *American Banker*, January 23, 2020, https://www.americanbanker.com/news/warren-presses-big-bank-ceos-on-climate-change-policies.

41. Elizabeth Warren, "Stop Wall Street from Financing the Climate Crisis" (plan released during Warren's 2020 presidential campaign), Warren Democrats, accessed October 3, 2022, https://elizabethwarren.com/plans/financing-the-climate-crisis.

42. David Roberts, "The Next President Can Force the Financial Sector to Take Climate Change Seriously," *Vox*, February 7, 2020, www.vox.com/energy-and-environment/2020/2/7/21127596/climate-change-financial-sector-dodd-frank-risk; Teresa Johnson, "Treat Climate Change as a Systemic Risk to Global Finance," *Financial Times*, March 9, 2020; Matt Hughes and Fernanda Borges Nogueira, "How Dodd-Frank Could Curb the Climate Crisis—Right Now," Roosevelt Institute blog, January 29, 2020, https://rooseveltinstitute.org/how-dodd-frank-could-curb-the-climate-crisis-right-now/.

43. Executive Order on Climate-Related Financial Risk, May 20, 2021, www.whitehouse.gov/briefing-room/presidential-actions/2021/05/20/executive-order-on-climate-related-financial-risk/.

44. Bank of America, "Bank of America Announces Actions to Achieve Net Zero Greenhouse Gas Emissions before 2050" (press release), February 11, 2021, https://newsroom.bankofamerica.com/.

45. "Tell Banks: Defund Line 3," Stop the Money Pipeline, accessed October 3, 2022, https://stopthemoneypipeline.com/defund-line-3/; Rainforest Action Network, *Who's Banking Line 3 and Keystone XL: Tar Sands Pipelines Being Rammed through in a Pandemic*, December 2020, www.ran.org/wp-content/uploads/2020/12/RAN-Briefing_Line3_KXL.pdf.

CONCLUSION

1. Matt Kempner, "Things to Know about Atlanta's Colonial Pipeline, Hit by Ransomware," *Atlanta Journal-Constitution*, May 10, 2021, www.ajc.com/news/. For a short history of Colonial Pipeline, see Barry Parker and Robin Hood, *Colonial Pipeline:*

Courage Passion Commitment (Chattanooga, Tenn.: Parker Hood Press, 2002); "$350 Million Pipeline to Have Wide Effect," *Washington Post*, April 4, 1962, B9; "Colonial Pipeline Worries Shipper," *Atlanta Constitution*, August 19, 1962, 46; "Pipeline Firm Here Plans Record Job," *Atlanta Constitution*, March 8, 1962; Clifford Krauss, "What to Make of the Pipeline Hack," *New York Times*, May 12, 2021, B1.

2. Kempner, "Things to Know."

3. "Colonial Pipeline to Write Check for Spill," *Daily Press* (Newport News, Va.), March 22, 1994, B1; House Committee on Public Works and Transportation, Subcommittee on Investigations and Oversight, *Colonial Pipeline Rupture*, 103 Cong., 1st sess., May 18, 1993; Parker and Hood, *Colonial Pipeline*, 60–61; National Transportation Safety Board, "Pipeline Accident Brief," Pipeline Accident Number DCA-98-MP-002, March 22, 1999, https://www.ntsb.gov/investigations/AccidentReports/Reports/PAB9901.pdf.

4. Dow Jones Newswires, "Colonial Pipeline Co. Is Sued by the U.S. for Oil Pipeline Spills," *Wall Street Journal*, November 29, 2000, C21; U.S. Department of Justice, "U.S. Reaches Landmark Settlement with Colonial Pipeline for Oil Spills in Five States" (press release), April 1, 2003, www.justice.gov/archive/opa/pr/2003/April/03_enrd_201.htm.

5. "Oil Edges Higher, Gas Slides a Little as Effects of the Break in the Colonial Pipeline Linger," *Wall Street Journal*, October 28, 1994, C12; Kempner, "Things to Know"; Parker and Hood, *Colonial Pipeline*, 59.

6. Myles McCormick and Derek Brower, "US Petrol Stations Emptied by Panic Buying after Pipeline Hack," *Financial Times*, May 11, 2021, www.ft.com.

7. Ed Bastian, "Delta CEO on Fuel Shortage's Impact on Airline Travel," video, interview by Craig Melvin, *Today*, May 12, 2021, www.today.com/video/; Kelly Yamanouchi, "Hartsfield-Jackson, Delta: Backup Jet Fuel Supplies Keep Planes Flying," *Atlanta Journal-Constitution*, May 13, 2021, www.ajc.com/news/.

8. Yamanouchi, "Backup Jet Fuel Supplies."

9. Meg Wagner et al., "What's Happening at US Gas Stations," CNN, May 13, 2021, www.cnn.com/business/live-news/us-gas-demand-05-12-21/index.html.

10. "Some Charlotte Businesses Report Impacts of Gas Outage," *Charlotte (N.C.) Observer*, May 12, 2021, www.charlotteobserver.com/news/.

11. "Company Structure and Facts," FedEx, accessed August 26, 2021, www.fedex.com/en-us/about/company-structure.html.

12. Christina Thompson, "Central Virginians Rushing to Fill Up over Gas Shortage Concerns," ABC 13 News, May 11, 2021, https://wset.com/news/local/; Holly Viers and Matthew Lane, "Gas Prices on the Rise after Colonial Pipeline Shutdown," *Times News* (Kingsport, Tenn.), May 13, 2021, https://www.timesnews.net/news/local-news/; Samantha Oller, "Is There Life after Wal-Mart for Murphy?" *CSP* magazine, May 3, 2016, https://www.cspdailynews.com/csp-magazine/there-life-after-murphy-wal-mart.

13. David Goldman, "Major Banks Hit with Biggest Cyberattacks in History," CNN Business, September 28, 2012, https://money.cnn.com/2012/09/27/technology/bank-cyberattacks/index.html; Matt Egan, "Hackers Paralyzed a Pipeline. Banks and Stock Exchanges Are Even Bigger Targets," CNN Business, May 12, 2021, www.cnn.com/2021/05/12/business/ransomware-attacks-banks-stock-exchanges/index.html; Jessica Bursztynsky, "Bank of America Spends over $1 Billion Per Year on Cybersecurity, CEO Brian Moynihan Says," CNBC, June 14, 2021, www.cnbc.com/2021/06/14/bank-of-america-spends-over-1-billion-per-year-on-cybersecurity.html.

14. America Hernandez and Laurens Cerulus, "What the US Colonial Pipeline Cyberattack Means for Europe," *Politico*, May 11, 2021, www.politico.eu/article/colonial-pipeline-us-cyberattack-europe-energy-infrastructure/; Barbara Bouldin, "NotPetya Holds Up a Stop Sign for FedEx," *United States Cybersecurity Magazine*, Spring 2018, www.uscybersecurity.net/csmag/notpetya-holds-up-a-stop-sign-for-fedex/.

15. 2021 Fortune 500 list, accessed August 24, 2021, https://fortune.com/fortune500; Karen Weise and Michael Corkery, "People Now Spend More at Amazon than at Walmart," *New York Times*, August 17, 2021, B1; Scott Neuman, "Jeff Bezos Completes His Blue Origin Flight to Space," NPR, July 20, 2021, https://www.npr.org/; Forrest Crellin, "Protestors Target Amazon in France Calling for Action on Climate Change," Reuters, July 2, 2019, https://www.reuters.com/.

16. 2021 Fortune 500 list, accessed August 24, 2021, https://fortune.com/fortune500. On ways Amazon built on Walmart's model, see Brad Stone, *The Everything Store: Jeff Bezos and the Age of Amazon* (New York: Little, Brown, 2013), 60–62, 72–75, 116–19, 125, 131–32, 161–62, 171, 245–46.

17. Jeff Bezos, "The Purpose of Going to Space," speech, Blue Origin event, Washington, D.C., May 9, 2019, in *Invent and Wander: The Collected Writings of Jeff Bezos* (Cambridge, Mass.: Harvard Business Review Press, 2020), 250.

18. IPCC, *Climate Change 2021: The Physical Science Basis*, accessed October 25, 2022, https://www.ipcc.ch/report/ar6/wg1/.

19. Bezos, "Purpose of Going to Space," 6, 247, 249.

Index